P9-AGA-363

IP 92 ~~40⁰⁰~~
22⁵⁰
lasers

understanding
LASER
technology

**An Intuitive Introduction
to Basic and Advanced Laser Concepts.**

By C. Breck Hitz

PennWell Books
PennWell Publishing Company
Tulsa, Oklahoma

Copyright © 1985 by
PennWell Publishing Company
1421 South Sheridan Road/P. O. Box 1260
Tulsa, Oklahoma 74101

Library of Congress Cataloging in Publication Data
Hitz, C. Breck
  Understanding laser technology.

  Bibliography: p.
  Includes index.
  1. Lasers. I. Title.
TA1675.H58 1984   621.36'6   84-9841
ISBN 0-87814-262-2

All rights reserved. No part of this book may be
reproduced, stored in a retrieval system, or
transcribed in any form or by any means, electronic
or mechanical, including photocopying and recording,
without the prior written permission of the publisher.

Printed in the United States of America

Contents

Preface

This book is intended to provide illuminating insight into the operation and application of lasers without requiring that the reader have a background in physics or mathematics. It has been my intention, both in preparing this text and in teaching laser-technology courses over the years, to emphasize the intuitive rather than the mathematical approach. Thus, I hope that the book will be useful to students studying to be laser technicians, to engineers and scientists in other fields wishing a quick introduction to lasers, and to nontechnical readers trying to understand some of the buzzwords of laser technology.

Understanding Laser Technology is organized into four sections. After the introductory Chapter 1, Chapters 2 through 5 discuss the basic concepts of light and optics, with special emphasis on laser light. The second major section of the book comprises Chapters 6 through 9 and presents the basic principles of laser operation. In many ways, these four chapters are the most fundamental in the book. Once the basic principles have been presented and discussed, the third section—Chapters 10 through 13—discuss techniques of altering the output from lasers. And the fourth section, Chapters 14 through 16, illustrate the principles presented in the rest of the book by discussing several specific lasers and laser applications. Of course, these chapters are by no means intended to be a complete discussion of laser applications. But they give the reader a flavor of how some of the unique properties of lasers can solve difficult problems.

The organizational similarity of this book and the excellent text on the same subject by O'Shea, Callen, and Rhodes is not coincidental because the present book is an outgrowth of a course I taught for several years from O'Shea. It has been my intention, though, that this book be more comprehensible and useful to students who lack a solid background in college physics and calculus.

A special comment about the chapter-end problems is appropriate. While some of these problems deal directly with the principles introduced in the text, others are intended to introduce more advanced concepts. Students may find that assistance from an instructor (or from this book's Instructor's Manual) is helpful in solving these problems.

It is an honor and a pleasure to express my gratitude to Professor Joel Falk of the University of Pittsburgh, not only for his careful review of the manuscript of this text but also for the many thoughtful discussions of quantum electronics we have had over the past fifteen years. I am also grateful to Professor Peter Milonni of the University of Arkansas, whose astute comments on the manuscript have been very helpful. Professor Anthony Siegman of Stanford University has made several very helpful suggestions about the subtleties of explaining quantum mechanics on an intuitive level. I would also like to acknowledge the suggestions and comments of many students who have taken my laser-technology course during the past ten years. And my appreciation for PennWell publisher Don Karecki and editor Kathryne Pile is enormous. Their long-suffering patience with this procrastinating author has finally paid off. Finally, the whole project would probably never have gotten started had it not been for the careful organization and typing of the original lecture notes by Dee Miller of Palo Alto.

C. Breck Hitz
October 1984

One

An Overview of Laser Technology

The laser is an elegant yet awesome tool that has provided a seemingly magic solution to numerous problems since its invention in 1960. Powerful lasers destroy airplanes in flight and slice through heavy steel as if it were cheese. More delicate and precise lasers, placed in the hands of skilled surgeons, can repair detached retinas in the human eye and stop bleeding deep inside a patient's body. Lasers read supermarket bar code labels in automatic cash registers, measure the distance to the moon or between terrestrial points with incredible accuracy, and may eventually drive fusion power plants to provide the human race with much of its energy during the next centuries. But what is a laser? How does it work? And what's so special about the light it produces?

The word *laser* is an acronym whose letters stand for "Light Amplification by Stimulated Emission of Radiation." As you'll see as you study this text, these words describe a process that generates an intense beam of light. There are several things in addition to its intensity that make the light in a laserbeam special. The light is very pure—that is, all the light rays in the beam are nearly the same color. And the light is extremely well collimated—that is, all the rays are headed in almost exactly the same direction. That means that the beam spreads out very little as it travels. These characteristics make the light in a laserbeam very special and allow lasers to be used in the applications described above.

What are the biggest applications of lasers? The answer to this question has to be based on a financial consideration: which laser applications result in the most spending? In the mid-1980s, the total spending on laser systems is running at something on the order of $4–$5 billion each year. How do the various applications of lasers add up to this total?

The largest worldwide market for laser systems belongs to laser printers, accounting for roughly one-quarter of the total mentioned above. Laser printers are faster and more versatile than other printing techniques, and the growth of the computer industry during this decade drives the need for fast, efficient printers.

Close behind laser printers come laser color separators, the second-largest market for laser systems. Whenever a color picture is mass printed—in a magazine, newspaper, or book, for example—it must be analyzed to determine the different colors of ink required to print the picture. Color separators that use lasers to record the results of this analysis have found widespread acceptance in the printing industry because of their speed and accuracy.

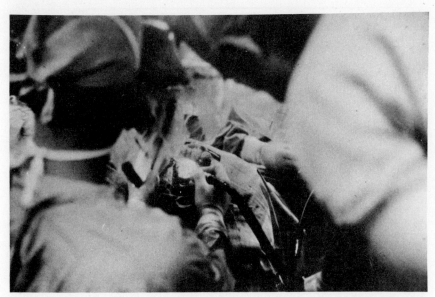

Fig. 1.1 *A laser is the brain surgeon's tool in this delicate operation* (photo courtesy Coherent Inc.)

Fig. 1.2 *Neodymium:YAG lasers are used as battlefield rangefinders and target designators* (photo courtesy Hughes Aircraft Co.)

Fig. 1.3 *Automated checkout stands in supermarkets use small helium-neon lasers* (photo courtesy Spectra-Physics Inc.)

Fig. 1.4 *This earthmover is accurately controlled by a laser whose beam is sensed by a detector mounted above the machine* (photo courtesy Spectra-Physics Inc.)

Optical communication is an application of equal importance to color separation. Fiberoptic cables can carry far more information in far less space than conventional electronic cabels can. Fiberoptics are being integrated into nearly every telephone system in the world, as well as into cable television networks and other communication systems. Lasers are often the light sources in optical communications systems.

Tactical military systems are another important application of lasers. Probably more than one-eighth of the total market for laser systems is for the rangefinders, target designators, and other tactical devices used by the armed services of all countries. (High-power laser weapons don't exist yet.) The tactical systems category includes target designators that pinpoint a target with laser light so "smart" ordnance can be directed to it and rangefinders that measure the distance to a target so conventional ordnance can be accurately aimed.

Smaller but still important applications of lasers include material processing, in which the energy of a laserbeam cuts, welds, or heat-treats material, and medical procedures, some of which were mentioned above. Lasers are used in the agricultural and construction industry to guide tractors and bulldozers. And lasers read the video-disks that many consumers buy to view on their televisions.

Research in physics and chemistry is yet another important application of lasers. Many of the discoveries of modern science could not have been made without lasers. Lasers scan, measure, and count things from incoming lumber in sawmills to newspapers rolling off printing presses. For that matter, lasers often make the plates used to print newspapers and set the type from which all manner of printing is generated. Lasers even find lighthearted applications in laser light shows and laser/holographic art.

This text is organized into four main sections. In chapters 2 through 5 we'll talk about laser light and what makes it so special. Next, in chapters 6 through 10 we'll see what goes on inside a laser to create this special kind of light. Then in chapters 11 through 13 we'll explain how lasers can be altered to change their output for special applications. The last three chapters—chapters 14 through 16—will take a look at applying some of the principles discussed in the first 13 chapters to real situations.

Fig. 1.5 *A high-power carbon dioxode laser cuts saw blades from a piece of thick steel* (photo courtesy Coherent Inc.)

Chapter

Two

The Nature of Light

What is light? How does it get from one place to another?

Those are the questions that will be addressed in this chapter. But the answers aren't all that easy. The nature of light is a difficult concept to grasp because light doesn't always act the same way. Sometimes it behaves as if it were composed of waves, and other times it behaves as if it were composed of particles. Let's take a look at how light waves act and at how light particles (photons) act, and then we'll discuss the duality of light.

Electromagnetic waves

Light is a *transverse electromagnetic* wave. Let's take that phrase apart and examine it one word at a time.

Fig. 2.1 is a schematic representation of a wave. It's a periodic undulation of something—maybe the surface of a pond, if it's a water wave—that moves with characteristic velocity, v. The wavelength, λ, is the length of one period, as shown in Fig. 2.1. The frequency of the wave is equal to the number of wavelengths that move past an observer in one second. It follows that the faster the wave moves—or the shorter its wavelength—the higher its frequency will be. Mathematically, the expression

$$f = v/\lambda$$

relates the velocity of any wave to its frequency, f, and wavelength.

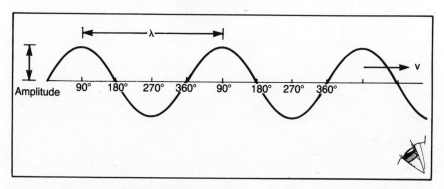

Fig. 2.1   *A wave and an observer*

The amplitude of the wave in Fig. 2.1 is its height, the distance from centerline to the peak of the wave. The phase of the wave refers to the particular part of the wave passing the observer. As shown in Fig. 2.1, the wave's phase is 90° when it is at its peak, 270° at the bottom of a valley, etc.

So much for "wave"; what does "transverse" mean? There are two kinds of waves: transverse and longitudinal. In a transverse wave, whatever is waving is doing so in a direction transverse (perpendicular) to the direction the wave is moving. A water wave is an example of a transverse wave because the thing that is waving (the surface of the water) is moving up and down, while the wave itself is moving horizontally across the surface. Ordinary sound, on the other hand, is an example of a longitudinal wave. When a sound wave propagates through air, the compressions and rarefactions are caused by gas molecules moving back and forth in the same direction that the wave is moving. Light is a transverse wave because the things that are waving—electric and magnetic fields—are doing so in a direction transverse to the direction of wave propagation.

Light is an "electromagnetic" wave because the things that are waving are electric and magnetic fields. A diagram of the fields of a light wave is shown in Fig. 2.2. It has an electric field (E) undulating in the

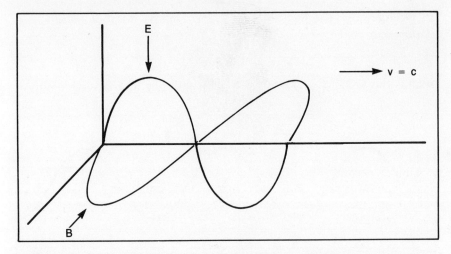

Fig. 2.2 *The electric (E) and magnetic (B) fields of a light wave*

vertical direction and a magnetic field (B) undulating in the horizontal direction. The wave can propagate through a vacuum because, unlike sound waves or water waves, it doesn't need a medium to support it. If the light wave is propagating in vacuum, it moves at a velocity c = 3.0 × 10^8 meters per second (m/s), the speed of light.[1]

Visible light is only a small portion of the electromagnetic spectrum diagramed in Fig. 2.3. Radio waves, light waves, and gamma rays are all transverse electromagnetic waves, differing only in their wavelength. But what a difference that is! Electromagnetic waves range from radio waves hundreds or thousands of meters in length down to gamma rays whose tiny wavelengths are on the order of 10^{-12} m. And the behavior of the waves in different portions of the electromagnetic spectrum varies radically, too.

But we're going to confine our attention to the "optical" portion of the spectrum, which usually means part of the infrared, the visible portion, and part of the ultraviolet. Specifically, laser technology is usually concerned with wavelengths between 10 microns (10^{-5} m) and 100 nanometers (10^{-7} m). The visible portion of the spectrum, roughly between 400 and 700 nanometers (nm), is shown across the bottom of Fig. 2.3.

---

[1]*It's convenient to remember that the speed of light is about 1 foot per nanosecond (ft/ns). Thus, when a laser produces a 3-ns pulse, the pulse is 3 ft in length.*

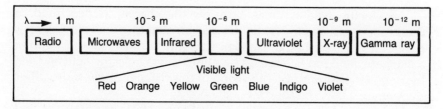

Fig. 2.3 *The electromagnetic spectrum*

The classical (i.e., nonquantum) behavior of light—and all other electromagnetic radiation—is completely described by an elegant set of four equations called Maxwell's equations, named after the nineteenth century Scottish physicist James Clerk Maxwell. Maxwell collected the conclusions of several other physicists, then modified and combined them to produce a unified theory of electromagnetic phenomena. His equations are among the most important in physics. Here's what they look like in the absence of dielectric or magnetic materials:

$$\nabla \cdot E = \rho$$

$$\nabla \cdot B = 0$$

$$\nabla \times E + \frac{\partial B}{\partial t} = 0$$

$$\nabla \times B = J + \frac{\partial E}{\partial t}$$

Now, these are differential equations, but you don't have to understand differential calculus to appreciate their simplicity and beauty.[2] The first one—Gauss's law for electricity—describes the shape of an electric field (E) created by electric charge (ρ). The second equation—Gauss's law for magnetism—describes the shape of a magnetic field (B) created by a magnet. The fact that the right side of this equation is zero means that it is impossible to have a magnetic monopole (for example, a north pole without a south pole.)

---

[2]$\nabla \cdot E$ is pronounced "divergence of E"; $\nabla \times E$ is pronounced "curl of E"; $\partial E/\partial t$ is pronounced "partial time derivative of E."

An electric field is created by electric charge, as described by Gauss's law, but an electric field is also created by a time-varying magnetic field, as described by Faraday's law (the third equation). Likewise, a magnetic field can be created by a time-varying electric field and also by an electric current, J.[3] The shape of this magnetic field is described by Ampere's law, the fourth equation.

The fame of these four little equations is well justified, for they govern all classical electrodynamics and their validity even extends into the realm of quantum and relativistic phenomena. We won't be dealing directly with Maxwell's equations any more in this book, but they've been included in our discussion to give you a glimpse at the elegance and simplicity of the basic laws that govern all classical electromagnetic phenomena.

There are two special shapes of light waves that merit description here. Both of these waves have distinctive wavefronts. A wavefront is a surface of constant phase. To see an example, look at the *plane wave* in Fig. 2.4. The surface sketched passes through the wave at its maximum. Because this surface that cuts through the wave at constant phase is a plane, the wave is a *plane wave*.

The second special shape is a *spherical wave,* and, as you might guess, it is a wave whose wavefronts are spheres. A cross-sectional slice through a spherical wave in Fig. 2.5 shows several wavefronts. A spherical wavefront is the three-dimensional analogy of the two-dimensional "ripple" wavefront produced if you drop a pebble into a pond of water. A spherical wave is similarly produced by a point source, but it spreads in all three dimensions.

Wave-particle duality

Let's do a thought experiment with water waves. Imagine a shallow pan of water three ft wide and seven ft long. Fig. 2.6 shows the waves that spread out in the pan if you strike the surface of the water rapidly at

---

[3]Because there aren't enough letters in the English (and Greek) alphabets to go around, some letters must serve double duty. For example, in Maxwell's equations E represents the electric-field vector, but elsewhere in this book it will stand for energy. In Maxwell's equations J represents an electric-current vector and B represents the magnetic-field vector, but elsewhere J is used as an abbreviation for joules and B for brightness. The letter f is used in this book to mean frequency and to designate the focal length of a lens. The letters and abbreviations used in this book are consistent with most current technical literature.

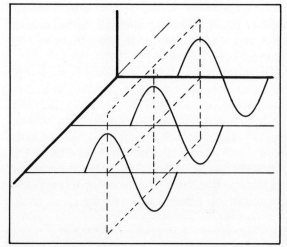

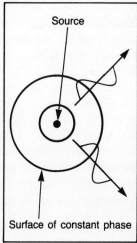

Fig. 2.4 *A plane wave*

Fig. 2.5 *A spherical wave*

point A. Now look at what happens at points X and Y. A wave crest will arrive at Y first because Y is closer to the source than X is. In fact, if you pick the size of the pan correctly, you can arrange for a crest to reach X just as a trough arrives at Y—and vice versa.

On the other hand, if you strike the water at point B, the wave crest will arrive at X first. But (assuming you're still using the correct-size pan) there will still always be a crest arriving at X just as a trough arrives at Y, and vice versa.

What happens if you strike the water at A and B simultaneously?

At point X, a crest from A will arrive at exactly the same time as a trough arrives from B. Likewise, a crest from B will be canceled out by a

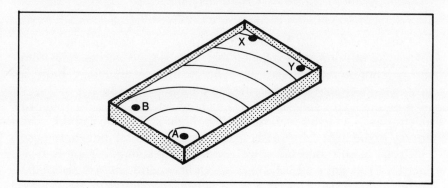

Fig. 2.6 *Wave experiment in a shallow pan of water*

trough from A. At point X, the surface of the water will be motionless. The same argument holds for point Y. But at a point halfway between X and Y, where crests from A and B arrive simultaneously, there will be twice as much motion as there was before.

A similar situation can be observed with light, as diagramed in Fig. 2.7. Here, two slits in a screen correspond to the sources, and dark stripes on a viewing screen correspond to motionless water at points X and Y. This experiment, called Young's double-slit experiment, will be analyzed in detail in Chapter 5. But here's the point for now: The only way to explain the observed results is to postulate that light is behaving as a wave. There is no possible way to explain the bright spot at the center of the screen if you assume that the light is made up of particles. However, it's easily explained if you assume light is a wave.

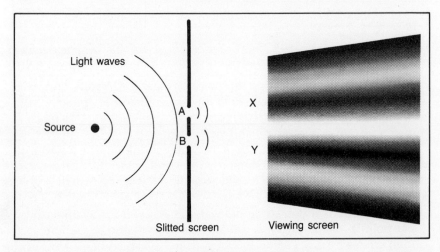

Fig. 2.7 *Optical analogy to wave experiment in Fig. 2.6.*

During most of the nineteenth century, physicists devised experiments like this one and explained their results quite successfully from the assumption that light is a wave. But near the turn of the century, a problem developed in explaining the photoelectric effect.

A photoelectric cell, shown schematically in Fig. 2.8, consists of two electrodes in an evacuated tube. When light strikes the cathode, the energy in the light can liberate electrons from the cathode, and these electrons can be collected at the anode. The resulting current is measured with an ammeter (A). This is a simple experiment to measure the current collected as a function of the voltage applied to the electrodes, and the data look like the plot in Fig. 2.9.

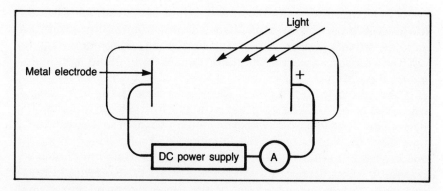

Fig. 2.8 *A photoelectric cell*

There is a lot of information in Fig. 2.9. The fact that current doesn't change with positive voltage (voltage that accelerates the electrons toward the anode) implies that every electron emitted from the cathode has at least some kinetic energy. An electron doesn't need any help to get to the anode. But as soon as the voltage starts to go negative, the current decreases. This implies that some of the electrons are emitted with very little energy; if they have to climb even a small voltage hill, they don't make it to the anode. The sharp cutoff of current implies that there's a definite maximum energy with which electrons are emitted from the cathode.

So some electrons are emitted with high energy, and some barely get out of the cathode. This makes sense if you assume that the high-

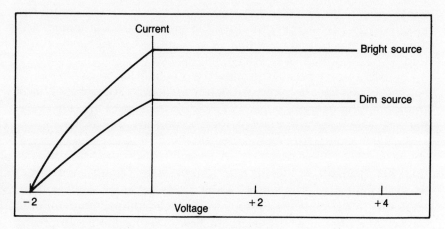

Fig. 2.9 *Current vs voltage for a photoelectric cell*

energy electrons came from near the surface of the cathode while the low-energy ones had to work their way out from farther inside the cathode. What's hard to understand is why the maximum energy for emitted electrons doesn't depend on the intensity of the light illuminating the cathode.

Think about it for a second. The electric field in the light wave is supposed to be exerting a force on the electrons in the cathode. The field vibrates the electrons, imparting energy to them so they can break free from the cathode. As the illumination intensifies—that is, the electric-field strength increases—the energy of vibration should increase. An electron right on the surface of the cathode should break free with more energy than it did before the illumination intensity was increased. In other words, the current from the bright source in Fig. 2.9 should go to zero at a greater negative voltage than it does from the dim source. But that's not what happens.

There are other problems. For example, you can easily figure out how much energy the most energetic electrons have when they leave the cathode. (For the experiment whose data appear in Fig. 2.9, those electrons would have two electron-volts of energy.) If all the energy falling on an atom can somehow be absorbed by one electron, how long does it take that electron to accumulate two electron-volts of energy?

The rate at which energy hits the whole surface is known from the illumination intensity. To calculate the rate at which energy hits a single atom, you have to know how big the atom is. Both now and back at the turn of the century when all this confusion was taking place, scientists knew that an atomic diameter was on the order of 10^{-8} centimeters (cm). For typical illumination intensities of a fraction of a microwatt per square centimeter, it takes a minute or two for an atom to absorb two electron-volts. But in the laboratory, the electrons appear immediately after the light is turned on with a delay of much less than a microsecond. How can they absorb energy that quickly?

In 1905, Albert Einstein proposed a solution to the dilemma. He suggested that light is composed of tiny particles called *photons*, each photon having energy

$$E = hf$$

where f is the frequency of the light, and h is Planck's constant (h = 6.63 × 10^{-34} joule-sec). This takes care of the problem of instantaneous electrons. If light hits the cathode in discrete particles, one atom can absorb one photon while several million of its neighbors absorb no energy. Thus, the electron from the atom that was hit can be liberated immediately.

Einstein's theory also explains why the maximum energy of electrons emitted from the cathode doesn't depend on illumination intensity. If each liberated electron has absorbed the energy of one photon, then the most energetic electrons (those that came right from on the surface of the cathode) will have energy almost equal to the photon energy. But increasing the illumination intensity means more photons, not more energy per photon. So a brighter source will result in more electrons but not more energy per electron. That's exactly the result shown in Fig. 2.9.

On the other hand, changing the color of light—that is, changing its wavelength and therefore its frequency—will change the energy per photon. In subsequent experiments, other physicists changed the color of light hitting the cathode of a photocell and observed data like those shown in Fig. 2.10. As the energy of the incident photons increases, so does the maximum energy of the photoelectrons.

So Einstein's photons explained the photoelectric effect and also explained other experiments that were conducted later and defied explanation from the wave theory. But what about experiments like Young's double-slit experiment, which absolutely can't be explained unless light behaves as a wave? How was it possible to resolve the seemingly hopeless contradiction?

The science of quantum mechanics developed during the early years of the twentieth century to explain this and other contradictions

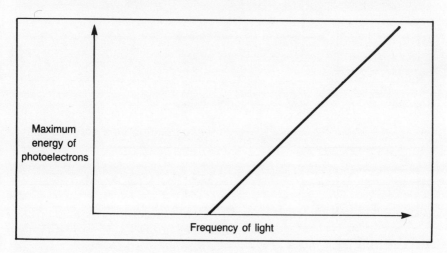

Fig. 2.10   *Energy of liberated electrons increases with photon energy*

in classical physics. Quantum mechanics predicts that when nature is operating on a very tiny scale—an atomic scale or smaller—it behaves much differently than it does on a normal, "people-sized" scale, so intuition has to be re-educated to be reliable on an atomic scale.

As a result of quantum mechanics, physicists now believe that the dual nature of light is not a contradiction. In fact, quantum mechanics predicts that particles also have a wave-like property, and experiments have proven that this property exists. By re-educating their intuitions to deal reliably with events on an atomic scale, physicists have found that the duality of light is not a contradiction of nature but is a manifestation of nature's extraordinary complexity.

If a laser produces a 1-ns, 1-joule (J) pulse of light whose wavelength is 1.06 micrometers (μm), there are two ways you can think of that light. As shown in Fig. 2.11, you can think of that pulse as a foot-long undulating electric and magnetic field. The period of the undulation is 1.06 μm, and the wave moves to the right at the speed of light. On the other hand, you could think of the laser pulse as a collection of photons, as shown in Fig. 2.12. All the photons are moving to the right at the speed of light, and each photon has energy $E = hf = hc/\lambda$.

Either way of thinking of the pulse is correct, provided that you realize neither way tells you exactly what the pulse is. Light is neither a wave nor a particle, but it's often convenient to think of light as one or the other in a particular situation. Sometimes light can act as both a

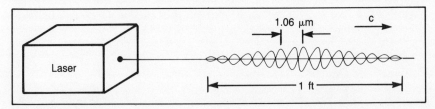

Fig. 2.11   *A 1-J, 1-ns pulse of 1.06-μ laser light pictured as a wave*

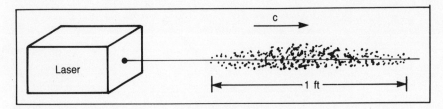

Fig. 2.12   *A 1-J, 1-ns pulse of 1.06-μ light pictured as photons*

wave and a particle simultaneously. For example, you could envision illuminating the cathode of a photocell with stripes of light from Young's experiment. Electrons would still be liberated instantaneously in the photocell, proving the particle-like nature of light despite the stripes, which prove light's wave-like nature.

Questions

1. What is the frequency of green light whose wavelength is $\lambda = 530$ nanometers (nm)? Roughly how many nanoseconds does it take this light to travel from one end of a 100-yd-long football field to the other?

2. Sketch Fig. 2.3 on a piece of paper. Beneath the figure, add the frequencies of the electromagnetic radiation that correspond to the wavelengths given above the figure.

3. A compass will not work properly underneath a high-voltage power line. Which of Maxwell's equations accounts for this? Which of Maxwell's equations describes the earth's magnetic field?

4. Calculate the frequency of the light wave emerging from the laser in Fig. 2.11. Calculate the number of photons emerging from the laser in Fig. 2.12.

≡Three

Refractive Index, Polarization, and Brightness

In Chapter 2 we talked about what light is. In the next several chapters we'll be talking about some of the properties of light. This chapter will begin with a discussion of how light propagates in a transparent medium like glass or water. Next we'll talk about the polarization of light. It's an important characteristic that has to do with the orientation of the electric and magnetic fields that make up the light wave. The chapter will conclude by discussing what is meant by the brightness of an optical source.

Light propagation–refractive index

The speed of light in a vacuum is 3×10^8 m/s, but it moves less rapidly in a transparent medium like glass or water. The electrons in the medium interact with the electric field in the light wave and slow it down. This reduction of velocity has many important consequences in the propagation of light. The *refractive index* of a material is determined by how much light slows down in propagating through it. The index is defined as the ratio of light's velocity in vacuum to its velocity in the medium. Here are the refractive indices for some common transparent materials:

Material	n
Dry air	1.0003
Crown glass	1.517
Diamond	2.419
Yttrium aluminum garnet (YAG)	1.825
Ice (−8°C)	1.31
Water (20°C)	1.33

The values listed above are approximate at best because the index of refraction of a material depends a little on the wavelength of light passing through it. That is, red light and blue light travel at exactly the same velocity in vacuum, but red light will travel a little faster in glass. This effect is called *dispersion*.

The velocity change that light experiences in moving from one medium to another accounts for the bending, or refraction, of light at the interface. Fig. 3.1 shows wavefronts passing through the interface. Consider wavefront AB. In (a) the light at both A and B is moving at c = 3×10^8 m/s. In (b), the light at B has entered the medium and has slowed down, while the light at A hasn't yet slowed. The wavefront is distorted as shown. In (c), the light at A has also entered the medium and the planar wavefront has been restored, propagating in a different direction than it had been outside the medium. If you think about the way a bulldozer turns, by slowing one tread relative to the other, you'll have a pretty good analogy.

Incidentally, now you can understand how a prism separates white light into its component colors. In Fig. 3.2, red and blue wavefronts approach the prism together. But because the blue light slows down a little more when it enters the glass, it is bent at a slightly greater angle than the red light. Thus, the two colors emerge from the prism at slightly different angles and will separate from each other as they travel away from the prism.

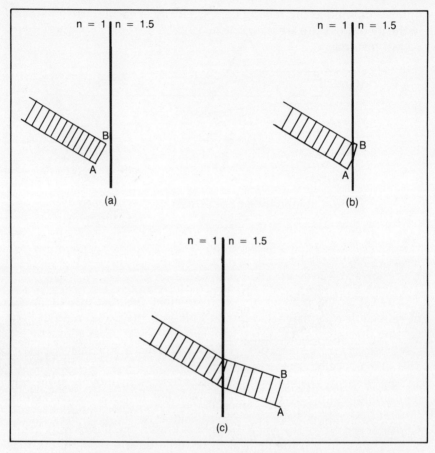

Fig. 3.1 *Refraction of a wave front at an interface between optical media*

The frequency of light is an absolute measure of the energy of the light. Because energy is conserved, the frequency of light cannot change as the light moves from one medium to another. But the wavelength depends on the velocity, according to the equation introduced in the beginning of Chapter 2:

$$\lambda = v/f$$

So when light moves from one refractive medium to another, its wavelength changes by an amount proportional to the ratio of refractive indices of the two media. It's analogous to what happens if a small child

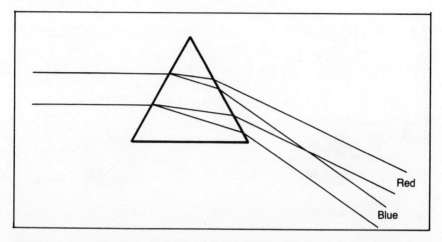

Fig. 3.2 *A prism refracts different colors at different angles because it is dispersive*

is bouncing up and down in the back seat of a car. Fig. 3.3a shows the path his nose will follow as the car moves along at 50 miles per hour (mph). On the other hand, if the child doesn't gain or lose any energy (that is, if he keeps bouncing at the same frequency), the path his nose follows will have half its former wavelength when the car slows to 25 mph, as shown in Fig. 3.3b. Likewise, when the speed of light decreases as it moves from one medium to another, the wavelength decreases by a proportional amount.

This wavelength-changing phenomenon gets more interesting if you consider dispersion. For example, take two light waves, one of which has twice the wavelength of the other in vacuum. As shown in Fig. 3.4, when these two waves enter an optical medium, their wavelengths will change. But because the refractive indices for the two waves are different, the fractional wavelength changes will not be the same, and the one wavelength will no longer be twice the other. As we'll see in Chapter 13, this effect has important consequences in nonlinear optics.

Polarization

Remember that light is composed of orthogonal electric and magnetic waves, as shown in Fig. 3.5. The polarization of light is the direction of oscillation of the electric field. For example, the light in Fig. 3.5 is plane polarized because the electric field oscillates only in one plane

(a)

(b)

Fig. 3.3 *The path followed by the child's nose at (a) 50 mph and (b) 25 mph. The wavelength is shorter as the car slows*

(the y-z plane). And since this plane is vertical, the light is vertically polarized. Horizontally polarized light is shown in Fig. 3.6.

Fig. 3.6 is the last time in this book that you'll see the magnetic field represented in a light wave. It's the electric field that determines the polarization of the light, and that's the only field we're going to be concerned with. Not that the magnetic field isn't important, because indeed it is. It wouldn't be light without the magnetic field. But as a matter of convenience, we're only going to show the electric field in future diagrams.

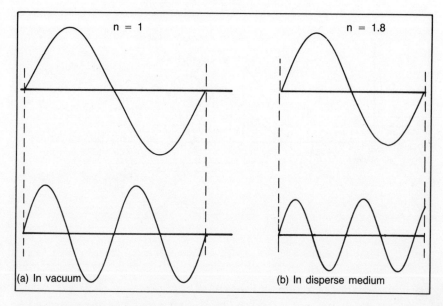

Fig. 3.4 *Although one wavelength is twice the other in vacuum, dispersion in a transparent medium destroys that relationship*

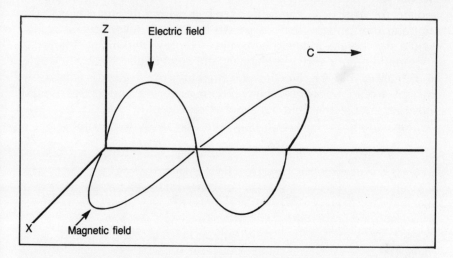

Fig. 3.5 *Light is composed of orthogonal electric and magnetic waves. This light is vertically polarized because the electric field oscillates in a vertical plane*

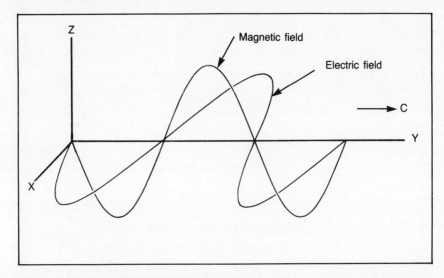

Fig. 3.6   *A horizontally polarized light wave*

The light that you're reading this book by is a collection of many waves: some polarized vertically, some horizontally, and some in between. The result is unpolarized light, light in which the electric field oscillates in all random directions.

Suppose the light wave in Fig. 3.6 were coming directly at you. What would you see as the wave hit your retina? At the beginning, you'd see no electric field at all. Then you'd see it growing and diminishing to the left and then to the right, as shown in Fig. 3.7. Over several cycles, you'd see the electric field behaving as represented in Fig. 3.9b.

On the other hand, unpolarized light coming directly at you would look like Fig. 3.8. The direction and amplitude of the electric field at any instant would be completely random. Over several cycles, you'd observe the electric field behaving as represented in Fig. 3.9c: a collection of random field vectors going off in random directions.

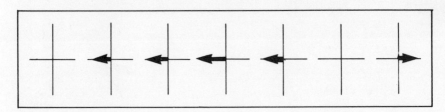

Fig. 3.7   *What the electric field of Fig. 3.6 looks like as it moves past you; each picture is an instant in time later*

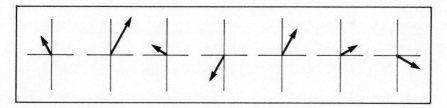

Fig. 3.8   *What the electric field of unpolarized light looks like as it moves past you*

There's another type of polarization that is important in laser optics. The polarization vector (which is the same as the electric-field vector) describes a circle as circularly polarized light moves toward you, as shown in Fig. 3.10. Over several cycles, you'd see the electric field behaving as represented in Figs. 3.11 and 3.12. Just as plane-polarized light can be vertically or horizontally polarized, circularly polarized light can be clockwise or counterclockwise polarized.

Polarization components

In order to understand how the polarization of light can be manipulated by devices like waveplates, Pockels cells, and birefringent filters, it's necessary to understand how light can be composed of two orthogonally polarized components.

Suppose at some point in space there are two electric fields, as shown in Fig. 3.13. These two fields are equivalent to a single electric

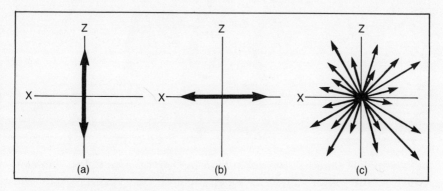

Fig. 3.9   *Orientation of the electric field for (a) vertically polarized light coming directly at you, (b) horizontally polarized light, and (c) unpolarized light*

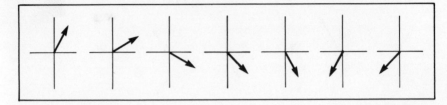

Fig. 3.10 *What the electric field of (clockwise) circularly polarized light looks like as it moves past you*

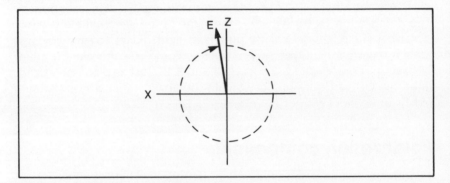

Fig. 3.11 *Circularly polarized light coming directly at you*

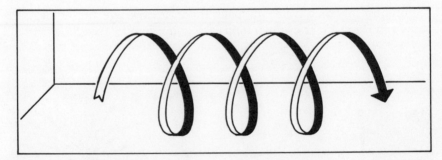

Fig. 3.12 *The path traced by the tip of the electrical field vector in circularly polarized light*

field, which is called the *vector sum* of the two original fields. You can construct the vector sum by lining up the individual vectors, tip to tail, without changing the direction they point (Fig. 3.14). It's meaningless to try to say whether it's the two original fields or their vector sum that "really" exist at that point in space; the two pictures are exactly equivalent. What's more, two different fields could describe the situation equally well if their vector sum is the same as that of the first two. Fig. 3.15 shows two such fields that might exist at the point. In other words, Fig. 3.13, Fig. 3.14, and Fig. 3.15 are three different ways of describing the same physical situation.

Now take a look at Fig. 3.16, which shows two electric fields. (Don't confuse this drawing with Fig. 3.6, which shows the electric and magnetic fields of a light wave. Fig. 3.16 shows the electric fields of two light waves.) At any point along the y-axis, the two fields are equivalent to their vector sum. And what does the vector sum look like as the waves of Fig. 3.16 move past you? That's shown in Fig. 3.17; the two waves in Fig. 3.16 are exactly equivalent to the single, plane-polarized light wave shown in Fig. 3.18. Fig. 3.16, Fig. 3.17, and Fig. 3.18 are different ways of describing exactly the same physical situation.

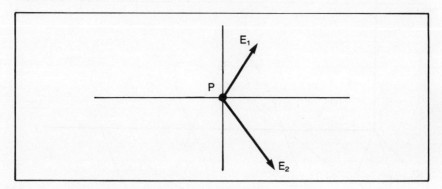

Fig. 3.13 *Two electric fields at a point in space*

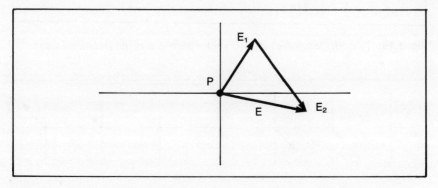

Fig. 3.14 *To find E, the vector sum of E_1 and E_2 in Fig. 3.13, line E_1 and E_2 up tip-to-tail without changing their direction*

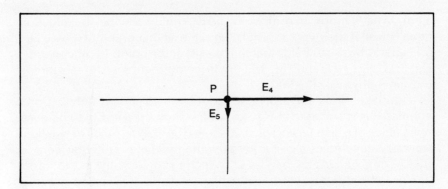

Fig. 3.15 *These two electric fields produce the same vector sum as E_1 and E_2 in Fig. 3.13*

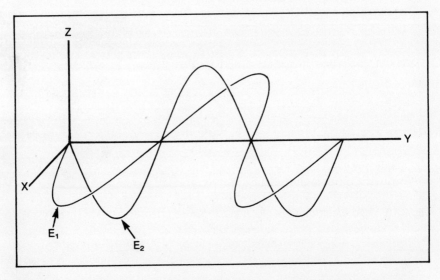

Fig. 3.16 *Two electric waves whose vector sum is a plane-polarized wave*

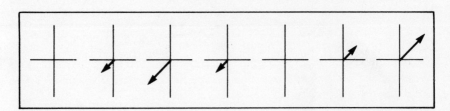

Fig. 3.17 *What the vector sum of the waves in Fig. 3.16 looks like as it moves past you*

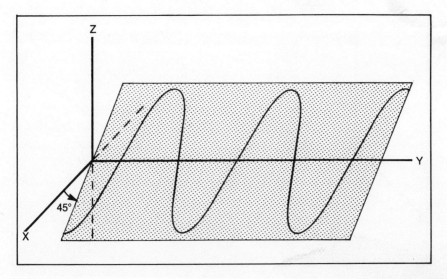

Fig. 3.18 *The two waves in Fig. 3.16 add at every point along the y-axis to produce a single field that looks like this*

If you go backward through this explanation, you'll see that any plane-polarized light wave can be thought of as the vector sum of two orthogonal components that are in phase with each other.

The phrase "in phase with each other" is crucial. What happens if the two orthogonal components are out of phase with each other, as in Fig. 3.19? Look at various points along the y-axis and see what the vector sum of the two waves looks like. Three such points are shown in Fig. 3.20. This light is circularly polarized.

So circularly polarized light can be thought of as the vector sum of two orthogonal components that are 90° out of phase with each other.

Now it gets really interesting: you can change the plane-polarized light to circularly polarized light by changing the phase between the orthogonal components. How? By using a *birefringent* device called a *quarter-wave plate*. A birefringent material has two refractive indices called the *ordinary* and *extraordinary* indices. Light in one polarization sees the ordinary index, while light in the orthogonal polarization sees the extraordinary index. Remember that the speed of light in a medium depends on the refractive index. If the two components of plane-polarized light shown in Fig. 3.16 travel through a birefringent material, they'll go through at different speeds—and they'll emerge out of phase, as shown in Fig. 3.19. The whole picture is shown in Fig. 3.21. Notice that the length of the birefringent material has to be selected

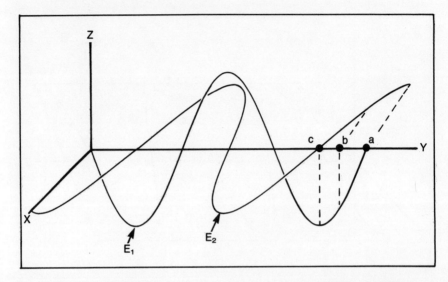

Fig. 3.19 *Two electric waves whose vector sum is a circularly polarized wave*

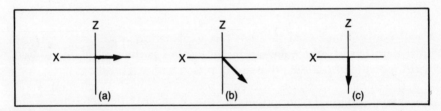

Fig. 3.20 *What the vector sum of the waves in Fig. 3.19 looks like as it moves past you*

carefully so the one polarization component is retarded exactly one-quarter wave with respect to the other.

Brewster's angle

If you've ever glanced at your reflection in a plate-glass window, you know that light is reflected from the interface between glass and air. That's because any discontinuity in refractive index causes a partial (or sometimes total) reflection of light passing across the discontinuity. The reflectivity of a glass-air interface for light at normal incidence is about 4%. So when you glance at your reflection in that plate-glass

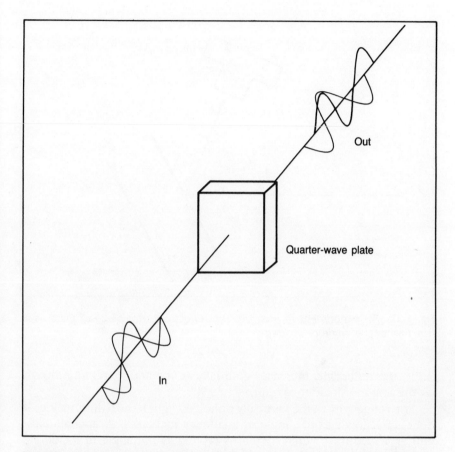

Fig. 3.21 *A quarter-wave plate changes linearly polarized light to circularly polarized light. It does so by retarding one component of polarization by a quarter wave with respect to the other*

window, you're seeing about 8% of the light incident on the window being reflected back to you—4% from each side of the glass.

The important point here is that the fraction of light reflected depends on the angle of incidence and the polarization of the light. Let's conduct a thought experiment to see how this works. If you set up a piece of glass and a power meter as shown in Fig. 3.22, you can measure the reflectivity as you rotate the piece of glass (and the power meter). What will your results look like? That's shown in Fig. 3.23, where the reflectivities for both horizontal and vertical polarizations are given as a function of the angle of incidence. For the horizontal polarization (that is, the polarization perpendicular to the plane of Fig.

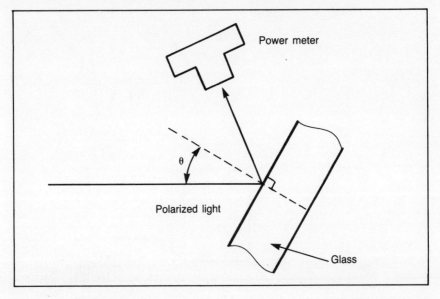

Fig. 3.22 *An experiment to measure the reflectivity of a piece of glass as a function of the angle of incidence*

3.21), the reflectivity increases gradually at first and then more rapidly as the angle of incidence approaches 90°.

The reflectivity of the vertically polarized light is more interesting. It decreases slowly as the angle of incidence increases, disappears altogether at about 57°, then increases rapidly as the angle of incidence approaches 90°. The angle at which the reflectivity disappears is Brewster's angle, and it depends on the refractive indices of the two media.[1]

A Brewster plate—a small piece of glass oriented at Brewster's angle—can be placed inside a laser to introduce a loss of about 30% (15% at each surface) to one polarization but no loss to the other polarization. This preferential treatment of one polarization is usually enough to make the laser lase in only that polarization.

Brightness

When you say that one light source is brighter than another, you mean that the brighter source creates a greater intensity on the surface

[1]*Brewster's angle is given by $\tan \theta_B = n_2/n_1$ where n_2 is the refractive index of the medium on the right if the ray travels from left to right.*

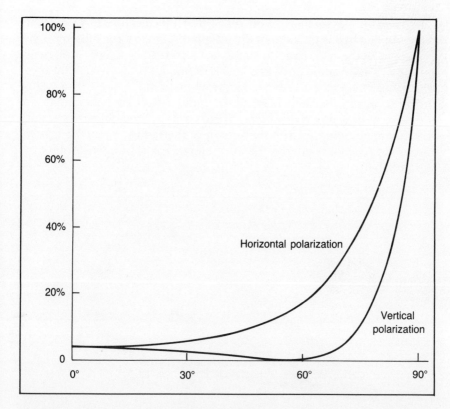

Fig. 3.23 *Reflectivity of the glass-air interface as a function of the angle of incidence*

of your retina when you look at the source. The intensity on this surface depends on the intensity of the source and the extent to which the light spreads out after it leaves the source. The faster light spreads out, the less reaches your eye. This spreading out of the light is called the *divergence* of the source, and it can be measured in terms of the solid angle formed by the light leaving the source.[2]

What's a solid angle? Well, you can think of a solid angle as the three-dimensional analogy to an ordinary plane angle. In general, a solid angle can have any irregular shape, like the one shown in Fig. 3.24. Or it can be regular, like an ice cream cone. The magnitude of a plane angle is measured in radians; the magnitude of a solid angle is

[2]*Because the beams from commercial lasers are usually symmetrical, divergence is more conveniently measured in plane rather than solid angles. Thus, the divergence of most lasers is specified in radians rather than steradians.*

measured in steradians. If you recall that the magnitude of a plane angle (in radians) is the ratio of the length of a circle's arc subtended by the angle to the radius of the circle, you can readily see how the magnitude of a solid angle (in steradians) is defined. It's the ratio of the area of a sphere's surface subtended by the angle to the square of the radius of the sphere. Why the square of the radius? Because solid angle is a measure of how much the angle spreads out in three dimensions, and that quantity increases with the square of the radius. Since the square of the radius is divided into the area, steradians are dimensionless (as are radians).

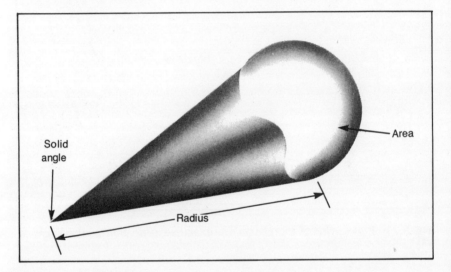

Fig. 3.24 *A solid angle*

In laser technology, the brightness of an optical source is defined as the source's intensity divided by the solid angle of its divergence:

$$B = \frac{P}{A\Omega}$$

where P is the power of the source, A is its cross-sectional area, and Ω is its divergence in steradians.

Notice that, because steradians are dimensionless, the dimensions of brightness are watts per square centimeter—the same dimensions as intensity.[3] But brightness is different from intensity because the intensity of a source doesn't depend on its divergence.

[3]The dimensions of brightness are often written W/cm²/sr where W is watts and sr is steradian—even though a steradian is not a real dimension.

Questions

1. There are 2π radians in a circle. How many steradians are there in a sphere?

2. Calculate the brightness of a one-watt source with a one-centimeter diameter that radiates into 10^{-6} steradian. Repeat the calculation if the source radiates into 2×10^{-6} steradian. Repeat the calculation if the source radiates into 10^{-6} steradian but its diameter is increased to 2 cm.

3. Polarizing sunglasses cut the glare reflected from horizontal surfaces (roads, etc.) better than ordinary sunglasses. Use the information in Fig. 3.21 to explain why this is so. Should the sunglasses reject horizontal or vertical polarization?

4. If you have two sets of polarizing sunglasses and hold them at right angles (so one blocks horizontal polarization and the other blocks vertical), how much light can get through both lenses? Now place the lens of a third pair between the first two "crossed" lenses. If the third lens is oriented to pass light polarized at 45° to the vertical, what do you see when you look through all three lenses? (Hint: "Nothing" is the wrong answer. Explain why.)

5. A half-wave plate retards one polarization component one-half wave with respect to the other. If a half-wave plate is substituted for the quarter-wave plate in Fig. 3.19, what is the polarization of the light on the other side of the plate? How does its polarization differ from the polarization of the input light?
    Suppose the light going into the half-wave plate is circularly polarized in the clockwise direction. What is the polarization of the light on the other side of the plate, and how does it differ from the input polarization?

6. The tip of the electric-field vector traces an ellipse (rather than a circle, as in Fig. 3.10) in elliptically polarized light. Explain how an elliptically polarized light can be resolved into two orthogonal components. What is the phase relation between the components?

7. Describe a fast and easy way, by observing the light reflected from a hand-held microscope slide, to figure out whether a laser beam is linearly polarized and, if it is, whether it's vertically or horizontally polarized.

Interference

The effects of optical interference can be observed with laser light or with ordinary (incoherent) light, but they can be explained only by postulating a wave-like nature of light. In this chapter we'll explore what optical interference is and note several examples of where it can be observed in everyday life. Then we'll examine two other examples that are especially important to laser technology. The first of these is Young's double-slit interference experiment, which embodies the principle of the acousto-optic effect as well as the whole field of holography. The second example will be a Fabry-Perot interferometer, which is an important instrument in its own right but is also very similar in principle to a laser resonator.

What is optical interference?

Two or more light waves *interfere* with each other when they're brought together in space to produce an electric field equal to the sum of the electric fields in the individual waves. The interference can be *destructive* if the waves are out of phase with each other or *constructive* if the waves are in phase. In Fig. 4.1a, the two waves that are brought together are in phase, so a bright spot—brighter than either of the

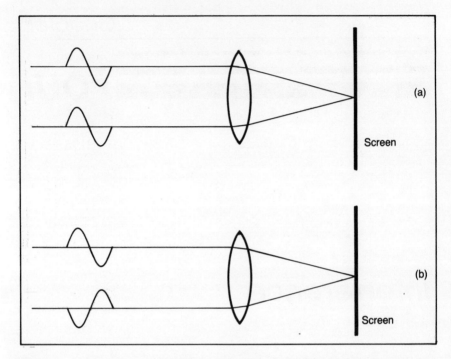

Fig. 4.1 *Constructive (a) and destructive (b) optical interference*

waves would produce by itself—is created on the screen. In Fig. 4.1b the two waves are out of phase, so they cast no illumination on the screen.[1]

As an example of how optical interference can occur, consider two smooth microscope slides separated at one end by a human hair, as shown in Fig. 4.2. An observer will see bright and dark bands if the slides are illuminated as shown because two waves will interfere at his retina. Fig. 4.3 shows where those two waves come from: one from the upper slide and one from the bottom. (Let's ignore reflections from the other two surfaces of the slides; they would behave the same way as the ones we're taking into consideration.) These two waves are brought together at the observer's retina by the lens of his eye.

---

[1]*The situations depicted in Fig. 4.1 can't exist by themselves because each one violates conservation of energy. The energy in the two waves in Fig. 4.1b doesn't just disappear, but in fact it turns up somewhere else in other waves from the same source.*

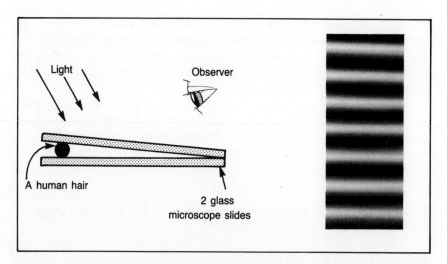

Fig. 4.2 *An example of optical interference. The observer sees evenly spaced bright and dark stripes, as shown on the right*

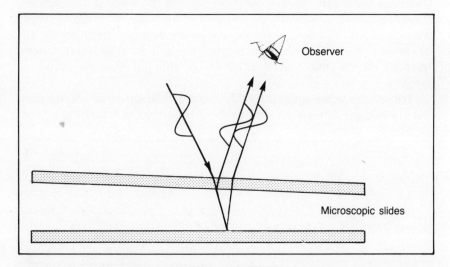

Fig. 4.3 *If the path difference for the two waves is an integral number of wavelengths, the observer sees constructive interference*

Now, these two waves started out being the same wave, but they were split apart by the partial reflection from the upper slide. The wave reflected from the bottom slide has a longer distance to travel before it reaches the observer's eye, and this "path difference" is important. If the path difference is an integral number of wavelengths, then constructive interference will occur at the observer's retina. On the other hand, if the path difference is an integral number of wavelengths plus one more half-wavelength, then the waves will be out of phase and will interfere destructively at the observer's retina.

Constructive interference is shown in Fig. 4.3. But if the observer shifts his gaze slightly, so that he's looking at a different place, the path difference will also change and destructive interference can occur. So as the observer shifts his gaze over the microscope slides, he sees a series of light and dark stripes as shown in Fig. 4.2.

Everyday examples of optical interference

If you've ever noticed the colors reflected from an oil slick on a puddle of water or reflected from the side of a soap bubble drifting through the air, you've seen optical interference from a thin film. Take the oil slick as an example. It's a thin film of oil floating on top of the water, as shown in Fig. 4.4. Sunlight striking the oil slick is reflected from the interface between oil and air and from the interface between water and oil. The two reflected waves are brought together at the observer's retina. Constructive interference will occur only if distances traveled by the two waves differ by an integral number of wavelengths.

For some wavelength (say, red), there will be constructive interference, while for other wavelengths the interference will be destructive.

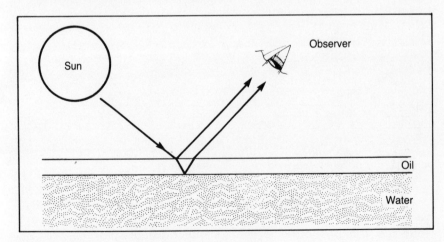

Fig. 4.4 *Optical interference from a thin film*

Thus, the observer sees red. When he shifts his gaze to another spot on the puddle, the path difference will change and he'll see a different color because there will now be an integral number of those wavelengths over the path difference.

It's the same idea for the soap bubble, except light is reflected from the inner and outer surfaces of the bubble.

If you hold a phonograph record up to the light so you're looking almost straight at its edge (tilt it just a little), you'll see a little rainbow of colors reflected from the grooves. That's because light glancing off different grooves follows different paths to your retina where it interferes constructively or destructively, depending on its wavelength.

The phonograph record acts like a diffraction grating, a device that will be discussed more in Chapter 11. Letters and stickers made of inexpensive diffraction gratings are often used to decorate automobiles. They're eye-catching because their color changes as the viewing angle changes. Sometimes jewelry is made from diffraction gratings for the same reason.

Young's double-slit experiment

Suppose a screen with two closely spaced slits is illuminated from behind with monochromatic light, as shown in Fig. 4.5. The pattern of light cast on the viewing screen is not what you might expect it would be. Instead of the two bright stripes shown in Fig. 4.6, many bright stripes appear on the viewing screen, as shown in Fig. 4.5. The brightest stripe is right at the center of the screen, midway between the two slits—where you'd expect the screen to be darkest.

This result, like any other interference effect, can be explained only by taking into account the wave-like nature of light. In Chapter 2 (Fig. 2.6), we learned how interference occurs between water waves from two sources, and the same logic will work with light waves. Fig. 4.7 shows the light wavefronts incident on the slits and the new wavefronts that emanate from each of the slits. Constructive interference occurs wherever a wave crest from one slit coincides with a crest from the other slit. The heavy lines in Fig. 4.7 show where constructive interference would cause a bright stripe to appear if a screen were placed at that location. The broken lines, on the other hand, show where the waves are exactly out of phase and would cast no illumination on a screen.

Fig. 4.8 shows a different way of looking at the same effect. In the figure, an incoming wave passes through the slits. The figure shows

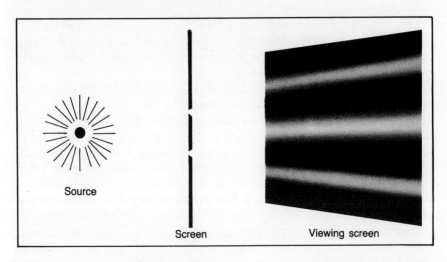

Fig. 4.5 *Young's double-slit experiment*

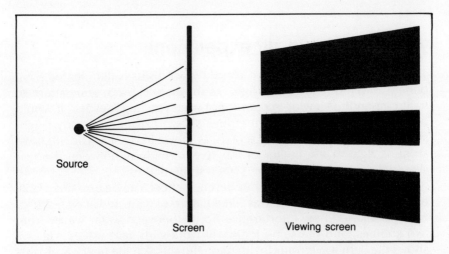

Fig. 4.6 *If light didn't behave as a wave, only two bright stripes would be produced on the viewing screen*

how that wave can arrive at two locations on the screen. At the center of the screen (P_1), the parts of the incoming wave that passed through the top and bottom slits have traveled the same distance and hence are in phase and interfere constructively to create a bright stripe. However, at a spot lower on the screen (P_2), the situation is different. Here the

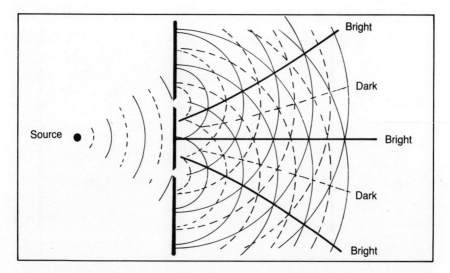

Fig. 4.7 *Alternating bright and dark bands are explained by the wave-like nature of light*

wave from the top slit has traveled farther than the other (it was on the outside of the curve) and has fallen behind. The two waves arrive at the screen out of phase, interfere destructively, and cast no light on the screen.

How much light is cast on the screen between P_1 and P_2? As you move from P_2 to P_1, the screen will become increasingly brighter because the two waves will be arriving more closely in phase. And any other point on the screen will be brightly illuminated, dimly illuminated, or not illuminated at all, depending on the relative phases of the two waves when they arrive at that point. The screen will look as it's pictured in Fig. 4.5.

Let's take a closer look at Fig. 4.8 and derive a mathematical expression for the angles where a bright stripe will be cast on the screen. If you assume that the distance from the slits to the viewing screen is much greater than the distance between the slits, then you can say that the paths taken by the two waves are approximately parallel. An enlargement of the part of the drawing showing the slits is presented in Fig. 4.9. Here, θ is the angle that the (almost-) parallel waves take when they leave the slits. The extra distance that the upper wave must travel to the screen is the distance labeled l in the figure. If this distance is an integral number of wavelengths, then a bright stripe will be cast on the

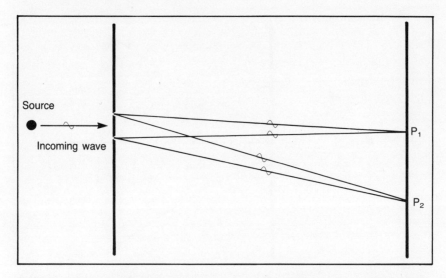

Fig. 4.8 *Screen illumination depends on the path difference traveled by light from each of the two slits*

screen at angle θ. But if you look at the triangle drawn in Fig. 4.9, you see that l = d sin θ, where d is the distance between the slits. Thus, the equation

$$n\lambda = d \sin \theta$$

defines the angles at which a bright stripe appears on the screen.

Fabry-Perot interferometer

Suppose a baseball team had two shortstops, each of whom had a .950 fielding average—that is, each shortstop stopped 95% of the balls hit in his direction. If the coach could play one shortstop in front of the other, then very few balls indeed would get past both athletes. In fact, unless one shortstop blocked the other's view, you could figure that only a couple of balls in a thousand (5% of 5%) would get through the pair.

Those two shortstops have a lot in common with a Fabry-Perot interferometer. If one 95% reflecting mirror is placed directly behind another, you might think that very little light indeed would get through both of them. However, that's not the case at all. As shown in Fig. 4.10, 100% of the incident light can be transmitted through the pair of mirrors.

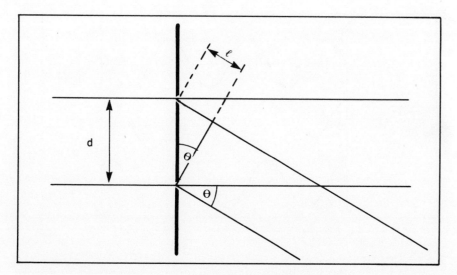

Fig. 4.9 *A closer look at Young's double-slit experiment*

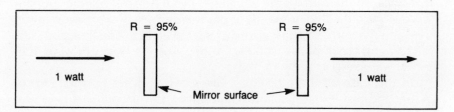

Fig. 4.10 *If a Fabry-Perot interferometer is resonant, it will transmit all the incident light, no matter how great the reflectivity of the individual mirrors*

The two mirrors in Fig. 4.10 form a Fabry-Perot interferometer, and it will transmit all the incident light if and only if it is *resonant*. To understand what this means, let's look more closely at what goes on between the mirrors.

If the first mirror (M_1) transmits 5% of the incident light and the second mirror (M_2) reflects 95% of this transmitted light, etc., some light will be trapped between the mirrors, bouncing back and forth. Several of these waves are shown in Fig. 4.11. (These waves bouncing back and forth actually all occupy the same volume in space, but in Fig. 4.11 they're shown separated for clarity.) Look at the phase relationship among all the waves traveling in one direction. In Fig. 4.11 these phases are random. The interferometer is nonresonant, and the mirrors act like shortstops, transmitting almost none of the incident light.

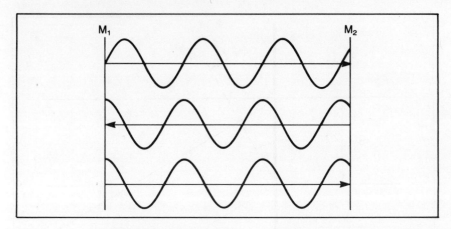

Fig. 4.11 *A nonresonant Fabry-Perot*

For a Fabry-Perot to be resonant, the separation between its mirrors must be equal to an integral number of half-wavelengths of the incident light. Such an interferometer is shown in Fig. 4.12. Notice here that because the mirrors are separated by an integral number of half-wavelengths, the light is exactly in phase with itself after one round trip between the mirrors. Thus, all the waves traveling in one direction (say, left to right) are in phase with each other. And the waves moving right to left will likewise all be in phase.

In this case all the individual waves between the mirrors add together and result in a substantial amount of power bouncing back and

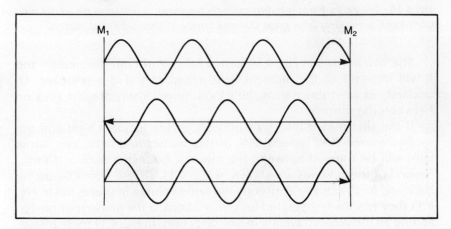

Fig. 4.12 *A resonant Fabry-Perot*

forth between the mirrors. For the interferometer shown in Fig. 4.10, about 20 watts will circulate between the mirrors, even though only one watt is incident on the interferometer. Now, those 20 watts are constantly reflecting off the second mirror, which transmits 5%. That's where the one watt of transmitted light, shown in Fig. 4.10, comes from.

Are you wondering about the light that travels right to left between the mirrors? It should be about 19 watts (because one of the 20 watts was transmitted through mirror M_2). What happens when this light strikes mirror M_1? M_1 is a 95% mirror, so is 5% of the 19 watts transmitted through it?

The answer has to be an emphatic "No." If 0.95 watt (5% of 19 watts) were transmitted, we'd have a device on our hands that transmits 0.95 watt to the left and 1.0 watt to the right, even though it's receiving only one watt. Thus, it would violate the law of conservation of energy.

But what happened to the 0.95 watt that should have been transmitted?

One way to understand what happens to the 0.95 watt that isn't transmitted through M_1 is to remember that there is one watt incident on the interferometer. This light hits M_1 from the left, and if M_1 were acting like a proper 95% mirror, it would reflect 0.95 watt back to the left. Now we have two "phantom" 0.95-watt beams coming from M_1: one transmitted through M_1 and one reflected from M_1. These two beams are exactly out of phase with each other and cancel each other. And just as the energy magically disappeared from the two beams in Fig. 4.1, the energy in the two "phantom" 0.95-watt beams from M_1 disappears . . . and shows up again in the light circulating between the mirrors.

So the input mirror M_1 doesn't act like a proper mirror when it's part of a Fabry-Perot interferometer because it doesn't reflect 95% of the incident light. The Fabry-Perot interferometer interacts with the electric and magnetic fields of the incident light as a single entity. It's misleading to try to figure out how one mirror by itself interacts with the light; you have to take the whole interferometer (both mirrors) into account.

Fig. 4.13 summarizes the behavior of a Fabry-Perot interferometer. If the interferometer is resonant (that is, if the spacing between its mirrors is equal to an integral number of half-wavelengths of the incident light), it will transmit the incident light, no matter how great the reflectivities of the individual mirrors. If the interferometer is nonresonant, it will reflect almost all of the incident light (assuming that the mirrors are highly reflective).

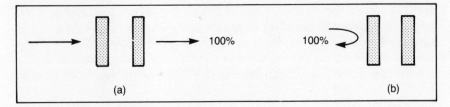

Fig. 4.13 *Almost all the incident light is transmitted through a resonant Fabry-Perot (a) and almost all is reflected from a nonresonant Fabry-Perot (b)*

A tunable source (for example, a dye laser) illuminates the Fabry-Perot interferometer in Fig. 4.14. As the wavelength of the source is changed, it will pass through several resonances with the interferometer, producing a series of transmission peaks as shown on the strip-chart recorder in the figure. Suppose that exactly 10,000 half-wavelengths of the source fit between the mirrors. Then the interferometer is resonant, and it transmits. But transmission ceases when the wavelength emitted by the source decreases somewhat because an integral number of half-wavelengths won't fit between the mirrors. Eventually, though, the wavelength will become short enough so that 10,001 half-wavelengths fit between the mirrors. Then the interferometer will be resonant once again.

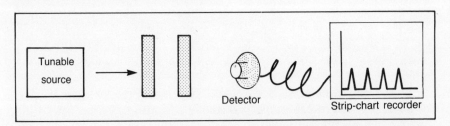

Fig. 4.14 *If light from a tunable source is incident of a Fabry-Perot, the interferometer will transmit whenever the incoming wavelength satisfies the resonance condition*

It's fairly simple to calculate the frequency difference between adjacent transmission peaks of a Fabry-Perot. The resonance requirement that the separation between the mirrors be equal to an integral number of half-wavelengths can be expressed mathematically as:

$$n\frac{\lambda}{2} = \ell$$

where λ is the wavelength of the light, ℓ is the mirror separation, and n is the integral number of half-wavelengths between the mirrors.

This equation can be solved for the wavelength of the resonant light:

$$\lambda = \frac{2\ell}{n}$$

Since $f = c/\lambda$ (see Chapter 1), the frequency of light at the nth resonance (that is, the resonance where there are exactly n half-wavelengths between the mirrors) is given by:

$$f_n = n \frac{c}{2\ell}$$

And it follows that the frequency of light at the $(n + 1)^{th}$ resonance is simply:

$$f_{n+1} = (n + 1) \frac{c}{2\ell}$$

Remember that we're finding the frequency difference between adjacent transmission peaks. To do that, we subtract the frequency of one from the frequency of its neighbor and find:

$$\Delta f = \frac{c}{2\ell}$$

The frequency separation of adjacent transmission peaks depends only on the spacing between the mirrors.

Questions

1. In a Young's double-slit experiment, calculate the distance from the central bright line on the viewing screen to the next bright line if the screen is one meter from the slits and the slits are 0.1 mm apart and illuminated with a 632.8-nm helium-neon (HeNe) laser.

2. Suppose a Fabry-Perot interferometer is constructed with mirrors that are highly reflective at both 488 nm and 532 nm. Calculate the

frequency separation between adjacent transmission peaks at each wavelength if the mirror separation is 1.000 cm. Are you surprised by the result? What would the frequency separation between adjacent transmission peaks be if the interferometer were used with a helium-neon laser at 632.8 nm?

3. If a Fabry-Perot interferometer is illuminated with a neodymium YAG laser whose nominal wavelength is 1.064 μm, what is the precise wavelength nearest to 1.064 μm that is transmitted? How many half-waves are there between the mirrors? Suppose the wavelength illuminating the interferometer is reduced until one more half-wave will fit between the mirrors. What is this new wavelength? How much does it differ from the previous wavelength? What is the frequency difference between the light transmitted in these two cases?

Chapter

Five

Laser Light

The previous chapters discussed some of the characteristics and behavior of light in general: the wave-like and particle-like nature of light, how it propagates through a dielectric medium, the polarization of light, and the phenomenon of optical interference. All these considerations apply to laser light as well. But laser light has some unique characteristics that don't appear in the light from other sources.

For example, laser light has far greater purity of color than the light from other sources. That is, all the light produced by a laser is almost exactly the same color, or *monochromatic*.

Another unique characteristic of laser light is its high degree of directionality. All the light waves produced by a laser leave the laser traveling in very nearly the same direction. One result of this directionality is that a laserbeam can be focused to a very small spot, greatly increasing its intensity.

These characteristics of monochromaticity and directionality, together with the phase consistency of laser light, are combined into a single descriptive term: *coherence*. Coherence is what makes laser light different from the light produced by any other source.

Monochromaticity

Because a glass prism is dispersive, it separates white light into its component colors (Fig. 5.1a). The *bandwidth* of white light is as wide as the whole visible spectrum, about 300 nanometers (nm). If light that is nominally red—maybe white light that is passed through a fairly good red filter—falls on the prism, it is separated into its component wavelengths, too. But in this case the bandwidth is far less, perhaps only 10 or 20 nm. The prism will produce a narrower band of colors, ranging from dark red to light red, as shown in Fig. 5.1b. But the prism will have no discernible effect on the red laser light in Fig. 5.1c because its bandwidth is vanishingly small compared to the red light from the filter in Fig. 5.1b. The bandwidth of a HeNe laser is typically somewhat less than 1 nm, and it can be reduced far below that amount by techniques described in Chapter 10.

It's important to note, though, that even a laser cannot be perfectly monochromatic. The light produced by a laser must have some non-zero bandwidth, even though that width is very slight by most standards. Why? Because a perfectly monochromatic optical source would

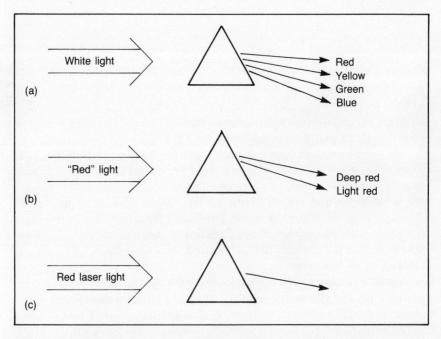

Fig. 5.1 *A prism can be used to understand the concept of monochromaticity*

violate the uncertainty principle, a foundation of modern physics. This principle holds that if you know (with no uncertainty) the wavelength of a source, you can know absolutely nothing about how long it has been on or how long it will stay on. That is, you have to suppose that it has been on forever and that it will remain on forever . . . clearly an impossible situation.

Directionality

Everyone has seen the publicity spotlights at circuses and used-car dealerships. Their beams of light penetrate the heavens, apparently diverging little as they disappear into the night sky. But while these spotlights produce beams that don't expand much over hundreds of yards, a laser beam the same size would propagate hundreds of miles without expanding very much.

The divergences of lasers are typically measured in milliradians. This very small divergence results from the requirement that light must make many round trips of the laser resonator before it emerges through the partially transmitting mirror. Only rays that are closely aligned with the resonator's centerline can make the required number of round trips, and these aligned rays diverge only slightly when they emerge (Fig. 5.2).

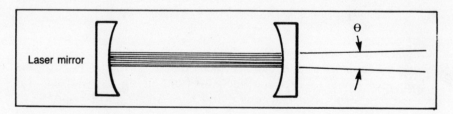

Fig. 5.2 *Because the light in a laser makes many round trips between the mirrors, it emerges with small divergence*

But just as it's impossible for a laser to be perfectly monochromatic, it's impossible for a laser (or anything else) to produce a nondiverging beam of light. Although the divergence of a laser beam can be very small when compared to light from other sources, there will always be some divergence. This is a basic property of light, called *diffraction*, which was explained in 1678 by the Dutch physicist Christian Huygens.

Huygens' principle lets you predict where a given wavefront will be later if you know where it is now. This is a useful thing to be able to do because it lets you understand how a light wave diverges. Simply stated, Huygens' principle holds that all points on the given wavefront can be considered sources that generate spherical secondary wavelets. The new position of the original wavefront is described by the surface of tangency to these wavelets.

To see how this works, let's look at the trivial case of a plane wave. Fig. 5.3 shows a plane wavefront and some of the points along that wavefront that can be considered sources of the secondary wavelets. These spherical secondary wavelets spread out as shown, and a short time later the new position of the wavefront can be deduced by constructing the dotted surface tangent to each wavelet.

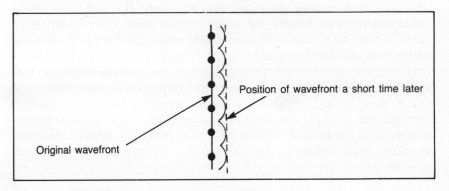

Position of wavefront a short time later

Original wavefront

Fig. 5.3 *Huygens' principle applied to a plane wave*

You may be asking yourself, "What happens to the wavefront that would be tangent to the back surfaces of the wavelets? Is Huygens trying to tell me there's another wavefront going backwards?" Of course there isn't, but Huygens didn't have a very good response to that question.

Usually, it's assumed that the intensity of the secondary wavelets diminishes to zero in the backward direction, thus getting rid of the backward-moving wavefront. Huygens' principle isn't a rigorous law of physics (after all, when Huygens formulated it in the seventeenth century, he had no idea what light really was), but it's often useful pedagogically. The only truly rigorous explanation of how light behaves depends on solving Maxwell's equations, but often we can gain some intuitive insight on a less-formal level.

As another example of Huygens' principle, Fig. 5.4 shows a spherical wave and some of the points on that wavefront that can be considered sources of secondary wavelets. These wavelets move away from

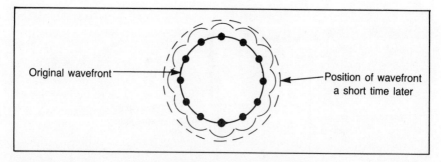

Original wavefront

Position of wavefront
a short time later

Fig. 5.4 *Huygens' principle applied to a spherical wave*

their sources, and a surface of tangency to them is the new spherical wavefront. That is, Huygens' principle predicts that the solid wavefront in Fig. 5.4 will develop into the wavefront represented by the broken line. As in the case with the plane wave, you must assume that the secondary wavelets diminish to zero in the backward direction.

Now let's use Huygens' principle to explain the phenomenon of diffraction. Fig. 5.5 shows a plane wave incident upon an aperture and some of the points that might produce secondary wavelets. These secondary wavelets spread out as shown in the figure, but some of them are blocked by the edges of the aperture. A surface of tangency to the remaining wavelets tends to wrap around at the edges as shown, so the light diverges after it has passed through the aperture.

If the aperture is much larger than the wavelength of the light passing through it, the divergence will be small. A small aperture, on the other hand, produces a large divergence. Mathematically, the full-angle divergence (in radians) is given by

$$\theta = 2.44 \, \frac{\lambda}{D}$$

where D is the diameter of the aperture.

But Fig. 5.5 doesn't tell the whole story because it doesn't take interference into account. If you let the light that has passed through the aperture in Fig. 5.5 fall onto a screen, you won't see a single bright spot. What you'll see is a diffraction pattern, as shown in Fig. 5.6. If you recall Young's double-slit experiment in Chapter 4, it's easy to understand where this pattern comes from. In Young's experiment, the light intensity at any point on the screen depends on how the waves from each slit add together at that point. For example, the point will be dark if the

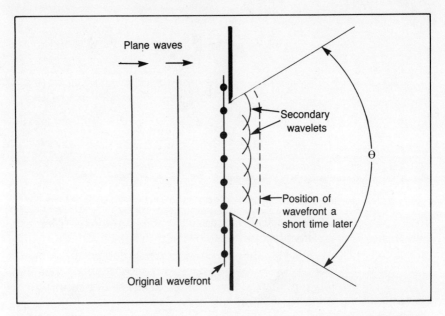

Fig. 5.5   *Divergence of an apertured plane wave, according to Huygens' principle*

waves add up out of phase with each other. In plane-wave diffraction, illustrated in Figs. 5.5 and 5.6, the intensity at any point on the screen depends on how all the secondary wavelets add together at that point. The result will be a central bright disk surrounded by light rings, or the diffraction pattern in Fig. 5.6. The equation above specifies the angle subtended by the central disk, which contains 84% of the light that passes through the aperture.

It's important to understand the significance of diffraction. For example, you can't reduce the divergence of light to an arbitrarily small angle by placing an aperture a great distance from a point source. Fig. 5.7 shows that the light which passes through the aperture will have a greater divergence than the geometrical divergence of the incident light if the aperture is small enough.

The divergence of a laser beam can be very small—even smaller than the divergence of plane waves diffracted through an aperture. Chapter 9 will discuss transverse modes and the many shapes that a laser beam can have, but for the moment let's be concerned only with Gaussian beams, which have intensity profiles as shown in Fig. 5.8. Mathematically, the intensity never vanishes completely, but conventionally the "edge" of the beam is taken to be the place where the intensity has dropped to about <u>14% (or $1/e^{-2}$)</u> of its maximum value.

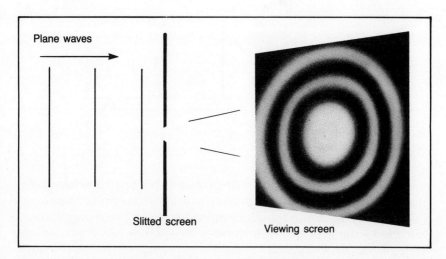

Fig. 5.6   *Diffraction of plane waves through an aperture*

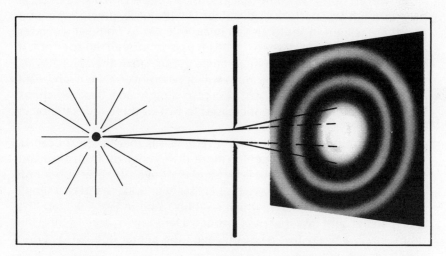

Fig. 5.7   *Diffraction makes the actual divergence of light greater than the geometrical divergence*

The divergence of a Gaussian beam (in radians) is given by the equation:

$$\theta = 1.27 \frac{\lambda}{D}$$

where D is the diameter of the beam from edge to edge at its smallest

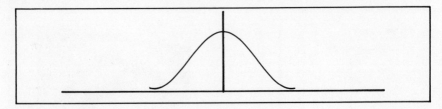

Fig. 5.8 *Intensity profile of a Gaussian beam; it's brightest at the center, dimmer toward the edges*

point, or at its "waist."[1] Notice that this is just about half the divergence of a plane wave passing through an aperture of diameter D. Moreover, a Gaussian beam doesn't produce diffraction rings like the light that is diffracted from a plane wave through an aperture. As it propagates through space, a Gaussian beam just expands—it doesn't change shape.

However, don't conclude that a Gaussian beam with diameter D can pass through an aperture of diameter D without ill effect. Remember that there is light in the Gaussian beam outside its nominal diameter, and this light is blocked when the beam passes through the aperture. In fact, the beam will diverge in an angle greater than the ideal $1.27\lambda/D$ radians, and faint diffraction rings will appear around the main beam. But these negative effects will decrease rapidly and become almost negligible as the aperture is increased to two or three times the beam diameter.

One benefit of a laser's small divergence is that the beam can be focused to a smaller spot than the more-divergent light from a conventional source. To see why this is true and to derive an expression for the diameter of the focused beam, it's necessary to understand two simple laws of optics. These laws, which are illustrated in Fig. 5.9, state that 1) all parallel light rays that pass through a lens are focused to the same point and 2) any ray that passes through the center of a lens is undeviated by the lens.

Fig. 5.10 shows a diverging beam focused to a small spot by a lens.[2] To calculate the diameter S of the focused spot, we'll add two more

---

[1]*The word "waist" has by convention come to mean the radius of the beam at its smallest point, not the diameter. If you know the waist of a beam, you must multiply it by two before using this equation.*

[2]*Fig. 5.10 shows the beam about the same size as the lens for clarity. In fact, the lens should be at least twice the diameter of the beam, lest it act as an aperture and introduce diffraction effects.*

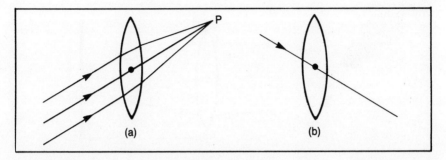

Fig. 5.9 *Two lens rules: (a) All parallel rays are focused to a common point. (b) Any ray through the center of the lens is undeviated*

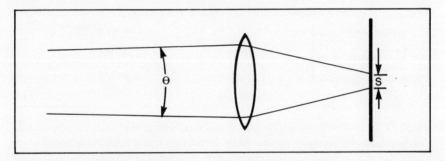

Fig. 5.10 *A beam of divergence θ is focused onto a screen*

rays, *c* and *d*, which are parallel to *b* and *a*, respectively, and which pass through the center of the lens (Fig. 5.11). Because rays *c* and *d* pass through the center of the lens, they are not bent. Since ray *b* is parallel to ray *c*, it will be focused to the same point on the screen as ray *c*. The same logic holds for rays *a* and *d*. The parallelism ensures that the angle between rays *c* and *d* is equal to the angle between rays *a* and *b*. But the angle θ, in radians, is given by θ = S/f (this can be deduced from Fig. 5.11) where f is the focal length of the lens. Thus, the diameter of the focused spot is S = θf. You can conclude that the smaller the divergence of the beam, the smaller the focused spot. And the smaller the focused spot, the greater the intensity of the light.

Coherence

Two waves in a laser beam are shown in Fig. 5.12. These waves illustrate the unique characteristics of laser light. They have very nearly (a) the same wavelength, (b) the same direction, and (c) the same

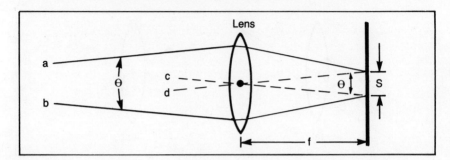

Fig. 5.11 *A construction used for calculating the diameter of the focused beam*

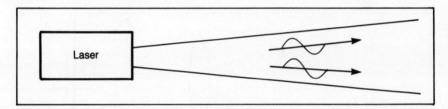

Fig. 5.12 *The waves of light from a laser are coherent; they all have very nearly (a) the same wavelength, (b) the same direction, and (c) the same phase*

phase. Together, these three properties make the light coherent, and this coherence is the property of laser light that distinguishes it from all other types of light.

All the things that can be done only with laser light can be done because laser light is coherent. Monochromatic laser light can be used to probe the structure of atoms or to control complex chemical reactions—because it's coherent. Highly directional laser light can transport energy over large distances or focus that energy to very high intensities—because it's coherent. Phase-consistent laser light can produce realistic three-dimensional holograms or create ultrashort pulses of light whose duration is only a dozen or so optical cycles—because it's coherent. The coherent light from a laser is indeed a different breed of light from that emitted by any other source.

What's more, the light from a laser exhibits both spatial coherence and temporal coherence. Its spatial coherence means that light at the top of the beam is coherent with light at the bottom of the beam. The farther you can move across the beam and still find coherent light, the greater the spatial coherence. Temporal coherence, on the other hand, comes about because two waves in a laserbeam remain coherent for a long time as they move past a given point. That is, they stay in phase with each other for many wavelengths. If you think about that for a minute, you'll see that the more monochromatic a laser is, the greater its temporal coherence.

Questions

1. Use Huygens' principle to explain the refraction (that is, the bending) of light at an interface between air and glass. (Hint: see Fig. 3.1)

2. What is the divergence of a HeNe beam whose waist (radius) is 1 mm?

3. U.S. astronauts have placed a laser retroreflector on the surface of the moon. If a Nd:YAG laser on earth is pointed at the moon, calculate the size of the beam when it reaches the moon. (The wavelength of Nd:YAG is 1.06 μm, and you may assume the beam has a 1-mm diameter at the laser and is not affected by the earth's atmosphere. The moon is roughly 250,000 miles from earth.) How would you modify the laser to make more of its light hit the retroreflector?

4. What phenomenon of wave propagation lets you hear around corners?

5. Explain why "temporal coherence" means the same thing as "monochromaticity."

6. Calculate the intensity (laser power/beam area) that would be created at your retina if you stared into the bore of a 0.5-milliwatt (mW) HeNe laser. Assume 1) the beam is significantly smaller than your pupil and has 0.5-milliradian divergence, 2) your retina is 2 cm from the lens of your eye, and 3) the lens has a 2-cm focal length. Compare this intensity to the intensity of the beam outside your eye.

7. One night, a careless technician pointed an argon-ion laser at a police helicopter flying overhead and temporarily blinded the pilot. (This is a true story. The pilot was able to land the helicopter safely, and the technician was subsequently arrested.) If the 5-watt laser were operating at a wavelength of 514 nm and a divergence of 5 milliradians and if the helicopter were 100 m directly overhead, cal-

culate the intensity (laser power/beam area) at the pilot's retina. Assume his night-accustomed pupil was 7 mm in diameter and that the retina of his eye was 2 cm from the lens, which had a focal length of 2 cm. For the purpose of calculation, assume the beam propagated like a Gaussian beam but had uniform intensity across its diameter. (Hint: the light hitting the pilot's eye diffracted through his pupil like a plane wave.)

Chapter

Atoms, Molecules, and Energy Levels

In Chapter 2 we saw that quantum mechanics is an explanation of nature that allows light to behave both as a wave and as a particle. But there are further implications of quantum mechanics: specifically, how it predicts energy is stored in atoms and molecules. The surprising— and far-reaching—conclusion is that energy can be added to or taken from an atom or molecule *only in discrete amounts*. That is, the energy stored in an atom or molecule is *quantized*. This means that while you might be able to add 1.27 or 1.31 electron-volts of energy to a particular atom, you can't add 1.26 or 1.28 or 1.30 electron-volts.[1] This is certainly a perplexing situation; it's like having a bucket that will hold 1.27 or 1.31 cups of water but can't hold 1.26 or 1.28 or 1.30 cups.

In this chapter we'll see how the requirement that energy be quantized affects the behavior of atoms and molecules, and in the next chapter we'll see that this behavior leads directly to laser action.

[1]*An electron-volt is a unit of energy.*

Atomic energy levels

Recall the basic structure of an atom. As shown in Fig. 6.1, it's a positively charged nucleus surrounded by a cloud of negative electrons, and each of these electrons moves in its own orbit around the nucleus. When energy is absorbed by the atom, the energy goes to the electrons. They move faster, or in different orbits. The crucial point is that only certain orbits are possible for a given electron, so the atom can absorb only certain amounts of energy. And once the atom has absorbed some energy, it can lose energy only in specified amounts because the electron can return only to allowed lower-energy orbits.

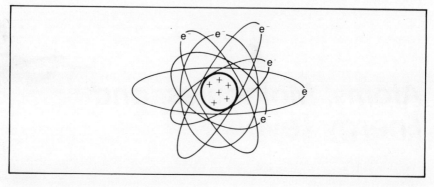

Fig. 6.1 *The positively charged nucleus of an atom is surrounded by an orbiting cloud of negative electrons*

The behavior of an atom can be shown schematically with an energy-level diagram like the one in Fig. 6.2. Here, the allowed energies for the atom are represented by different levels on the diagram. An atom in the ground state has energy E_0, while an atom in the first excited state has energy E_1, and so on. The atom loses energy $E = E_1 - E_0$ when it moves from level 1 to level 0. But an atom in level 1 cannot lose any other amount of energy; it must either keep all its energy or lose an amount equal to $E_1 - E_0$ all at once.

On the other hand, an atom in the ground state (level 0) can absorb only certain allowed amounts of energy. For example, the ground-state atom of Fig. 6.2 could absorb $E_1 - E_0$ and move to the first excited state (level 1), or it could absorb $E_2 - E_0$ and move to the second excited state, and so on. But the atom cannot absorb an amount of energy less than $E_1 - E_0$, nor can it absorb an amount of energy between $E_2 - E_0$ and $E_1 - E_0$.

This bizarre behavior on an atomic scale, the *quantization* of energy, is one of the fundamental results of quantum mechanics. As we discovered in Chapter 2, quantum mechanics explains that nature behaves differently on an atomic scale than it does on a "people-sized" scale. The theory seems bizarre to us because our intuition is based on our experience with nature on a people-sized scale. But the validity of quantum mechanics has been proven in many experiments that make the atomic-scale behavior of nature show up in the real, people-sized world.

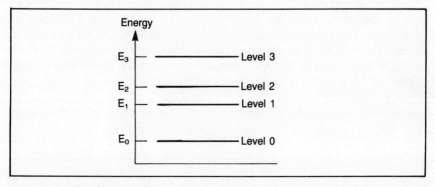

Fig. 6.2   *The allowed energy levels for an atom correspond to different orbital configurations of its electrons*

One of the ways an atom can gain energy is to absorb a photon. But the atom must absorb a whole photon because partial absorption is not (usually) allowed. That means that the energy of the photon must correspond exactly to the energy difference between two levels of the atom. For example, the ground-state atom in Fig. 6.2 could absorb a photon of energy $E_1 - E_0$ and move to the first excited state.

Because the energy of a photon is $E = hc/\lambda$, there's a restriction on the wavelength of light that can be absorbed by a given molecule. For the atom of Fig. 6.2, light of wavelength $\lambda = hc/(E_0 - E_1)$ will be absorbed and will boost the atom to its first excited state. But light whose wavelength doesn't correspond to an energy-level difference in the atom won't be absorbed. Red glass is red because it contains atoms (or molecules) that absorb photons of blue light but can't absorb red light. Ordinary glass has no atoms (or molecules) in energy levels that are separated from other levels by the amount of energy in visible photons.

Spontaneous emission and stimulated emission

There are several ways an atom in an excited state can lose its energy. The energy can be transferred to other atoms, or it can be emitted as light. If it is emitted as light, the wavelength of the emitted light will correspond to the energy lost by the atom. But there are two mechanisms by which the light can be emitted: spontaneous emission and stimulated emission.

The absorption and subsequent spontaneous emission of a photon is shown in Fig. 6.3. First, a photon whose energy is exactly right to boost the atom from its ground state to its first excited state is absorbed. The average atom will stay in this excited state for a period of time known as the *spontaneous lifetime*, which is characteristic of the particular transition. Many atomic transitions have spontaneous lifetimes of nanoseconds or microseconds, although much longer and shorter lifetimes are known. But eventually the atom will spontaneously emit the photon and return to its ground state. As indicated in Fig. 6.3, the photon is emitted in a random direction.

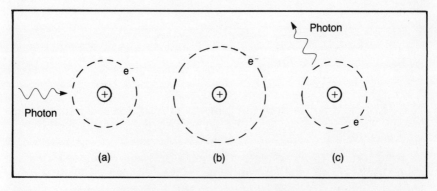

Fig. 6.3 *An atom absorbs a photon (a), which excites it for awhile (b). Later, the photon is spontaneously emitted (c)*

The process of stimulated emission is shown in Fig. 6.4. A second photon—one with exactly the same energy as the absorbed photon—interacts with the excited atom and stimulates it to emit a photon. Interestingly, this emission can take place long before the spontaneous lifetime has elapsed. The second photon isn't absorbed by the atom, but its mere presence causes the atom to emit a photon. As indicated in the figure, the light is emitted in the direction defined by the stimulating photon, so both photons leave traveling in the same direction. Of course, since the stimulating photon has the same energy as the emitted photon, the emitted light has the same wavelength as the stimulating light. The polarization of the emitted light is also the same as that of the stimulating light. Moreover, the emitted light has the same phase as the stimulating light: the peaks and valleys of the electromagnetic waves are all lined up with each other.

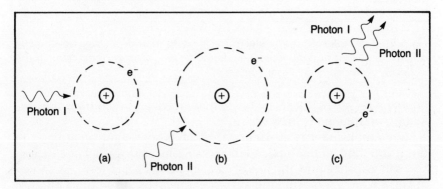

Fig. 6.4   *A second photon can stimulate the atom to emit in a time shorter than the spontaneous lifetime*

As we'll see later, stimulated emission is crucial to laser action. Indeed, the word *laser* is an acronym whose third and fourth letters stand for "stimulated emission." And, as we'll see in Chapter 7, the light emitted by a laser is coherent because its waves are all traveling in the same direction, all have the same wavelength, and are all in phase with each other.

Molecular energy levels

As you know, a molecule is composed of two or more atoms. Fig. 6.5 shows a simple molecule of only two atoms in which some electrons stay with their original nuclei and one electron is shared by both nuclei. Because a molecule is more complex than an atom, it has more types of energy levels than an atom does. In fact, there are three types of energy levels possible in a molecule: electronic levels, vibrational levels, and rotational levels.

A molecule can have electronic energy levels that are exactly analogous to the electronic energy levels of an atom. That is, the molecule's electrons move to more-energetic orbits when the molecule absorbs

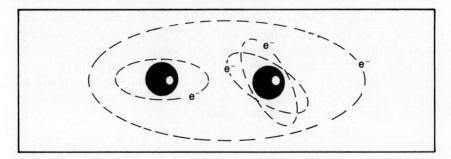

Fig. 6.5 *In a molecule, some electrons stay with their original nucleus, while others are shared or completely transferred to the other nucleus. (Not all the molecule's electrons are shown here)*

energy. Of course, these levels are quantized, and each transition has its own spontaneous lifetime.

But a molecule can also vibrate, which is something an atom can't do. If you think of the force holding the molecule together as a spring, you can visualize how the nuclei can vibrate back and forth, as shown in Fig. 6.6. So a molecule can absorb energy—by absorbing a photon, for example—and the absorbed energy can turn into vibrational energy in the molecule.

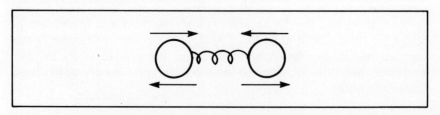

Fig. 6.6 *A molecule can store energy by vibrating*

The vibrational energy of a molecule is quantized, just as its electron energy is. That means that a molecule can't absorb just any amount of energy, but it can only absorb enough energy to move it from one allowed energy level to another. So you can draw an energy-level diagram for a molecule's vibrational levels just as you can draw a diagram for its electronic levels.

Finally, a molecule can absorb energy and start rotating about its axis, as shown in Fig. 6.7. An atom cannot store energy this way because, unlike a molecule, an atom's mass is completely symmetric.

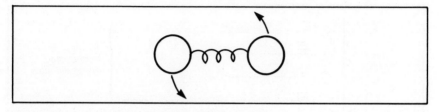

Fig. 6.7   *A molecule can store energy by rotating*

The rotational energy a molecule can possess is quantized so that a molecule can absorb or emit only the exact amounts of energy corresponding to a transition between allowed levels.

How do these three types of energy levels compare to each other? In general, transitions between electronic levels are the most energetic, and transitions between rotational levels are the least energetic. This conclusion is implicit in the table below, which summarizes the three types of energy levels and the wavelengths of transitions for each.

Level	Atoms	Molecules	Approximate λ of Most Transitions
Electronic	yes	yes	visible or ultraviolet
Vibrational	no	yes	near infrared
Rotational	no	yes	far infrared

Recall that the energy in a photon is inversely proportional to its wavelength, and you'll see that electronic transitions in general absorb or emit a greater amount of energy than vibrational or rotational transitions. And vibrational transitions in general involve more energy than rotational transitions. (There are, of course, exceptions to these rules.)

An energy-level diagram for a hypothetical molecule shows all three types of levels in Fig. 6.8. The closest-spaced levels in this simplified diagram are the rotational ones, while the electronic levels are farthest apart.[2]

---

[2]*The energy-level diagram for most real molecules would be more complex than Fig. 6.8, with many overlapping levels. For example, the higher rotational levels of the ground vibrational state might be more energetic than the ground rotational level of the first excited vibrational state.*

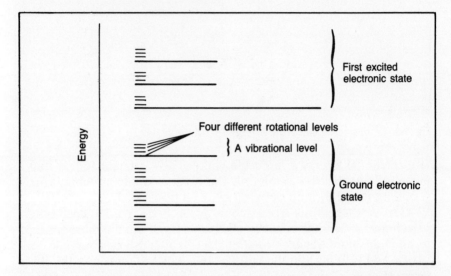

Fig. 6.8   *Electronic, vibrational, and rotational levels for a hypothetical molecule*

Because a molecule has so many different energy levels, its spectrum can be much more detailed than an atom's. For example, Fig. 6.8 shows only two electronic levels, and an atom with only two electronic levels would have only one emission or absorption line. But the hypothetical molecule in Fig. 6.8 could emit and absorb light at dozens of different wavelengths, corresponding to transitions involving each of many of the pairs of levels in the figure. In discussing the carbon dioxide laser in Chapter 14, we'll see that not every transition you could imagine is possible; selection rules arising from the conservation laws prohibit many transitions.

Some subtle refinements

The foregoing explanation of how energy is stored in atoms and molecules has deliberately trod roughly on some of the niceties of quantum theory in order to create a simple and somewhat intuitive model. And although this simple model is adequate to understand most of the principles of laser technology, you should know that it's an approximation that isn't quite correct from a theorist's point of view. In this section we'll discuss several of these subtleties of quantum mechanics.

The picture of an atom composed of marble-like electrons orbiting around a nucleus isn't consistent with modern quantum theory. The uncertainty principle says that the electrons are more like a negatively charged cloud surrounding the nucleus, not distinct particles. When the atom absorbs energy, the shape of this cloud changes to accommodate the extra energy.

In introducing the concept of spontaneous decay, we discovered that an excited atom will stay excited for a period of time known as its spontaneous lifetime. The concept of spontaneous lifetime is valid for a collection of atoms, in the sense that they will decay from the excited state with an average time constant equal to the spontaneous lifetime. But one particular atom can stay in an excited state for a longer or shorter time than the spontaneous lifetime. You must think of an average atom staying excited for its spontaneous lifetime in the same sense that you think of the average American family having 1.8 children.

But on an even more subtle level, the concept of an atom's being in one energy level and then moving abruptly to another energy level as it absorbs or emits a photon is incorrect. According to quantum theory, a single atom exists in a number of different energy levels simultaneously. When you perform a measurement of the atom's energy, you force the atom to have the amount of energy you measure at the instant you perform the measurement. But in general, the only correct description of an atom's energy is a description of the probability of making different measurements: "If I measure the energy in this atom, there's a 7% chance I'll find 2 electron-volts, a 20% chance I'll find 1.4 electron-volts, and a 73% chance I'll find 1 electron-volt." (Remember, quantum mechanics is not supposed to make perfect sense on an intuitive level until you have spent several years watching how it behaves— by taking a graduate degree in quantum physics, for example.)

How does the atom described above interact with light? That is, how does rigorous quantum mechanics explain optical absorption and stimulated emission? Quantum mechanics says that if the photon energy of the light (given by $E = hc/\lambda$) is equal to the difference between the possible measured energies of the atom, the light will cause the atom's energy-measurement probabilities to change. Specifically, the probabilities of finding the atom in either of the two energy levels separated by the light's photon energy tend to become equal.

Suppose the atom described above is irradiated with light whose photon energy is 1 electron-volt (i.e., light whose wavelength is about 1.2 μm). Initially, the atom has 7% probability of having 2 electron-volts and 73% probability of having 1 electron-volt. But if you made the measurement after the atom had been exposed to the 1.2-μm light for a while, you'd find that the probabilities were becoming equal. If you repeated the experiment 100 times (exposing the atom to the light for the same amount of time each instance), you might find the atom had 2 electron-volts 30 times, 1 electron-volt 50 times, and 1.4 electron-volts in the remaining 20 measurements. That is, you'd find that the probability of the atom's having 2 electron-volts had increased from 7% to 30%, and that its probability of having 1 electron-volt had decreased from 73% to 50%, as shown in Fig. 6.9.[3]

If you left the atom of Fig. 6.9 exposed to the light long enough, its probabilities would reach new equilibrium values. If the light were intense enough, the new equilibrium values would be equal (that is, a 40% probability of finding 2 electron-volts and a 40% probability of

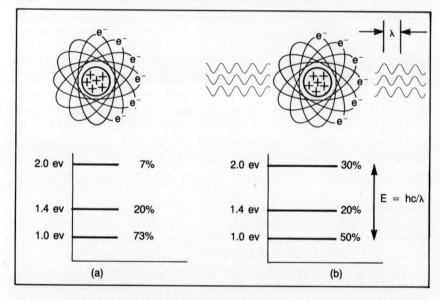

Fig. 6.9 *An atom might initially have 73% probability of being in the ground state, 20% probability of being in the first excited state, and 7% probability of being in the second excited state (a). When the atom is exposed to light whose photon energy (E − hc/λ) corresponds to the energy difference between the ground state and second excited state, the probabilities of finding the atom in those states start to change (b)*

[3]*If you had irradiated the atom with light whose photon energy didn't correspond to any transitions in the atom—say, light whose photon energy were 0.7 electron-volts—nothing would happen. The light wouldn't interact with the atom, and the probabilities would remain unchanged.*

finding 1 electron-volt). In this case, the transition is *saturated* because the probabilities don't change, no matter how much more light you pump into the atom.

In the situation depicted in Fig. 6.9, the atom ends up with more energy than it started with. Energy has been transferred from the light to the atom, so Fig. 6.9 represents the quantum mechanical version of optical absorption.

Fig. 6.10 shows the quantum mechanical explanation of stimulated emission. If the atom from Fig. 6.9b is immediately exposed to light whose photon energy is 0.6 electron-volt, the probabilities for the first and second excited states will change, and that change will tend to make the probabilities equal. That is, the probability of finding the atom in the second excited state might decrease from 30% to 28%, and the probability of finding it in the first excited state might increase from 20% to 22%. In this case, energy has been transferred from the atom to the light, so the light has been amplified by stimulated emission.

If the probabilities of Fig. 6.9a are the equilibrium values for the atom under a particular set of conditions, then the situation depicted in Fig. 6.10a is a nonequilibrium condition, and the probabilities will auto-

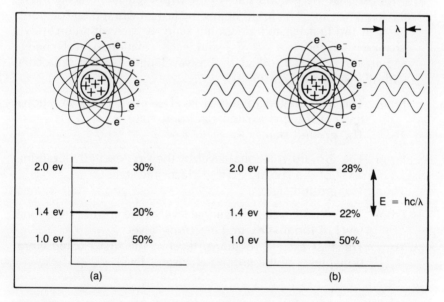

Fig. 6.10 *If the atom in Fig. 6.9b is exposed to light whose photon energy corresponds to the energy difference between the first and second excited states, the probabilities of finding the atom in either of these two states tend to become equal. In (b) the atom is shown shortly after the exposure has begun— before the probabilities have become equal*

matically drift back to those of Fig. 6.9a. The amount of time it takes them to drift back is determined by the spontaneous lifetime of the level. The energy given off by the atom as it drifts back may be in the form of spontaneous (optical) emission, it may be heat, or it may be collisionally transferred to the other atoms.

The picture of an atom absorbing or emitting a photon and instantaneously changing energy levels as it does so is not wrong. In fact, it's a very useful model. But be aware that the model is a simplification, and that the true quantum mechanical model is closer to that described in this section.

Questions

1. The 1.2-μm light in Fig. 6.9b has a photon energy of about 1 electron-volt. Calculate the photon energy in joules and from your result calculate the approximate conversion for converting electron-volts to joules. What is the wavelength of the light in Fig. 6.10b?

2. Suppose that the ground state of the hypothetical molecule in Fig. 6.8 lies 3.1×10^{-19} joules (J) below the lowest vibrational-rotational level of the first excited electronic state. If each vibrational level change requires 4×10^{-20} J and each rotational level change requires 5×10^{-21} J, calculate the wavelength of light associated with the following transitions.

From: The first excited rotational level of the ground vibrational level of the first excited electronic state
To: The ground state

From: The ground rotational level of the first excited vibrational level of the second excited electronic state
To: The ground state

From: The second excited rotational level of the ground vibrational level of the first excited electronic level
To: The first excited rotational level of the first excited vibrational level of the ground electronic level

Seven

Energy Distributions and Laser Action

In the previous chapter, we saw how energy is stored in atoms and molecules. It's stored in discrete amounts, and an atom can be thought of as making a transition from one energy level to another as it absorbs or emits energy.

In Chapter 6 we limited our attention to one atom or molecule at a time. This chapter will be different. Here, we'll examine the behavior of a collection of atoms, and we'll look at how the energy in a collection of atoms is divided among the individual atoms. That is, we'll find the answer to this question: If a jar holds 100 atoms of the same element, how many of these atoms are in the ground state? How many are in the first excited state? And so on.

And we'll see that, for an unusual type of energy distribution among a collection of atoms (or molecules), it's possible for light to be amplified as it passes through the collection. This amplification is the basis of laser action, and understanding it is absolutely crucial to understanding the operation of a laser.

This chapter will conclude by examining the two energy-level schemes of common lasers. Some of the mechanisms for pumping energy into common lasers will also be explained.

Boltzmann distribution

Suppose you have a sealed jar that contains 100 atoms of some element, as shown figuratively in Fig. 7.1. (Of course, a realistic-sized jar normally holds more like 10^{16} or 10^{20} atoms, but 10^2 is easier to work with.) In addition to the 100 atoms in the jar, there's also some energy in there. The first question you might ask yourself is, "How does the energy manifest itself?" That is, "What form does the energy take? How does it show up?"

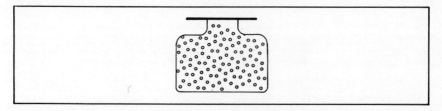

Fig. 7.1 *A jar containing exactly 100 atoms of an element*

Let's limit our discussion to the thermal energy in your jar. The more thermal energy in the jar, the higher the temperature will be. There are two ways that thermal energy in a collection of atoms manifests itself. Some of the thermal energy in your jar will show up in the motion of the atoms themselves: They carom around inside the jar, bouncing off the walls and each other. And some of the energy shows up in the electronic energy levels of the atoms. Thermal energy will boost some atoms to the first excited state, some to the second excited state, and so forth. But how many get to each excited state?

Boltzmann's Law is one of the fundamental laws of thermodynamics, and it dictates the population of each energy level if the atoms in the jar are in thermal equilibrium. Fig. 7.2 shows schematically the prediction of Boltzmann's Law.[1] In this figure the length of the bar representing each level is proportional to the population of that level. You

---

[1]*Mathematically, Boltzmann's Law predicts that the population of any energy level is related to the population of the ground level by the equation:*

$$N_i = N_0 \exp - [E_i/k_BT]$$

where N_i is the population of the ith level whose energy above ground level is E_i, and N_0 is the population of the ground level. The symbol k_B is Boltzmann's constant, 1.38×10^{-23} joules per degrees Kelvin, and T is the temperature. The results of this complicated-looking equation turn out to be fairly simple, as shown in Figs. 7.2, 7.3, and 7.4.

can draw an important conclusion immediately: No energy level will ever have a greater population than that of any level beneath it. Level E_3 is populated by more atoms than levels E_4, E_5, or E_6, but on the other hand it has fewer than levels E_2 or E_1.

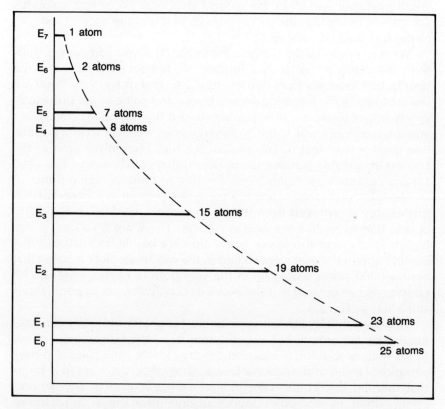

Fig. 7.2   *If the atoms in the jar are in thermal equilibrium, Boltzmann's law predicts they will be distributed as shown here, with increasing populations in the lower energy levels*

Remember that the distribution shown in Fig. 7.2 is an equilibrium distribution. That means that it's the normal way the atoms in the jar behave. It is possible to create an abnormal distribution for a short time (more about that later). But as long as the temperature of the jar doesn't change, the atoms will eventually come back to the distribution of Fig. 7.2.

Now you know that the picture of a jar containing so many atoms in a certain energy level isn't quite accurate because an atom in general doesn't really exist in only one energy level. The truly correct way to describe the jar in Fig. 7.2 is to say that each atom has a 25% probability of being in the ground state if you measured it, a 23% probability of being in the first excited state, and so on through the higher excited states. But that means if you take the trouble to measure all 100 atoms, you'll probably find 25 in the ground state, etc. So in that sense it's acceptable to say that the jar contains 25 atoms in the ground state, 23 in the first excited state, etc.

What happens to the distribution in the jar if you add energy to it? Suppose you put the jar in a furnace. Its temperature rises and the distribution changes from that of Fig. 7.2 to that of Fig. 7.3. There are fewer atoms in the low-lying energy levels, and some of the previously empty upper levels are now populated. But the important conclusion you drew earlier is still valid: No energy level will ever have a population greater than that of any level beneath it. Even in the limit of the highest imaginable positive temperature, that conclusion will be valid for any collection of atoms (or molecules) in thermal equilibrium.

Next, let's ask what happens if you place your jar in a freezer. Thermal energy is removed from the jar, and you'll wind up with a distribution that looks like the one in Fig. 7.4. There are no atoms in the higher levels, and the lower levels are very highly populated. What would happen if the jar were chilled all the way to absolute zero? In that case, all 100 atoms would be in the very bottom energy level, E_0. Of course, you've noticed that the previous conclusion about populations still holds in Fig. 7.4.

The same Boltzmann Law logic holds if your jar holds 100 molecules instead of 100 atoms, but the situation is a little more complicated. Recall that, in addition to electronic energy levels, molecules also have vibrational levels and rotational levels. Altogether, there are four forms that thermal energy can take in a jar of molecules: as translational energy when the molecules bounce around inside the jar, as rotational energy in the molecules, as vibrational energy in the molecules, and as electronic energy in the molecules. Raw energy flows back and forth among these four modes, keeping them in equilibrium with each other. Moreover, Boltzmann's Law governs the distribution of energy among the molecules. That is, the diagrams in Figs. 7.2, 7.3, and 7.4 could equally well represent the populations of rotational energy levels of a collection of molecules at three different temperatures.

Population inversion

As long as a collection of atoms is in thermal equilibrium, energy will be partitioned among them according to Boltzmann's Law. But it's

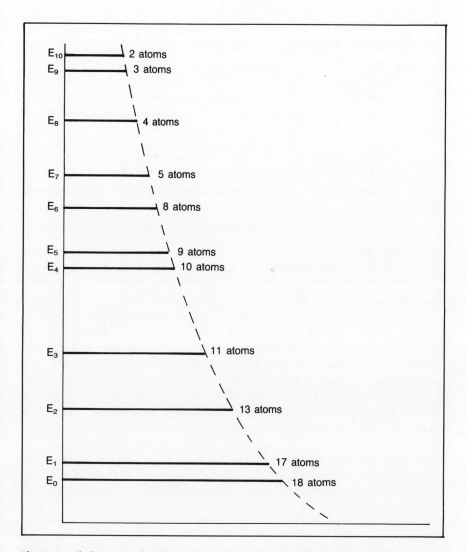

Fig. 7.3 *If the jar is heated, the atoms will redistribute themselves among energy levels as shown, but there will still always be increasing populations in the lower energy levels*

possible to create a collection of atoms that is not in thermal equilibrium. The collection won't stay in that nonequilibrium condition for very long, but for a short period you'll have a collection of atoms that violates the conclusion you drew in the previous section. For example, suppose you somehow plucked seven of the E_0 atoms out of the jar whose distribution is diagramed in Fig. 7.2. Then, for an instant at least,

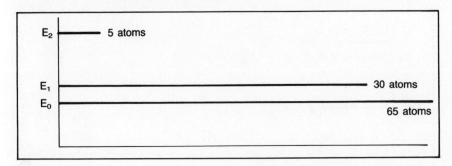

Fig. 7.4 *When the jar is chilled, a new equilibrium distribution results, with less total energy*

you'd have a distribution like that of Fig. 7.5. This distribution shows a *population inversion* between the E_1 and E_0 levels because the equilibrium populations are inverted: there are more atoms in level E_1 than in level E_0.

And there's more than one way to create a population inversion. Suppose you added some energy to the jar, but instead of adding random thermal energy, you added energy in very precise amounts. For example, you might think of setting your jar down in front of a gun shooting out a beam of electrons, each electron having the same velocity (and therefore the same energy) as all the other electrons. When one of the electrons collided with one of the 100 atoms in the jar, it could transfer its energy to the atom. Now, suppose the energy of the electrons coming from the gun were exactly equal to the energy difference between the E_2 level and the E_3 level. If one of these electrons collided with an atom in the E_2 level, it could excite that atom to the E_3 level. If the gun shot out electrons fast enough, it could pump atoms up to E_3 faster than they could decay spontaneously from that level. So you could create a population inversion between the E_3 and E_2 levels, as shown in Fig. 7.6.

The population inversions in Figs. 7.5 and 7.6 are nonequilibrium distributions—ones that won't last very long. If the electron gun in the previous example is turned off, the atoms in the jar will quickly revert to the distribution of Fig. 7.2. But there's nothing terribly unnatural about a nonequilibrium situation. You don't need to understand the intricacies of atomic energy levels to understand a nonequilibrium situation. Fig. 7.7a is another example of a nonequilibrium situation. The system of Fig. 7.7 is shown in its equilibrium condition in Fig. 7.7b. To create the nonequilibrium situation again, energy is input to the system

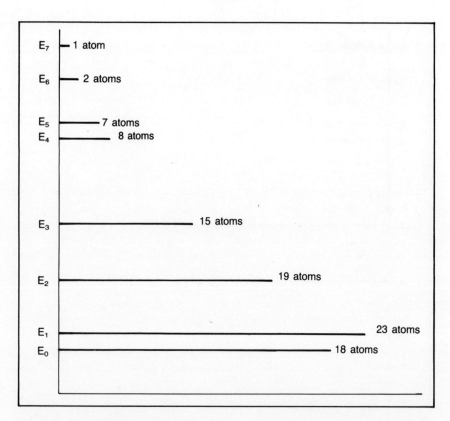

Fig. 7.5 *A population inversion could be created by plucking some E_0 atoms from the jar described in Fig. 7.2*

from an external source, as shown in Fig. 7.7c. Do you understand how the waterwheel and sun of this example are analogous to the jar of atoms and the electron gun of the previous example?

L.A.S.E.R.

In Chapter 6 we discussed stimulated emission, and in this chapter we've discussed population inversions. These are the two concepts necessary to understand the fundamental principle of the laser. As you know, the letters in the word *laser* stand for Light Amplification by Stimulated Emission of Radiation.

Go back to the hypothetical jarful of atoms we've been discussing. For a moment let's take all but one of the atoms out of the jar. Suppose

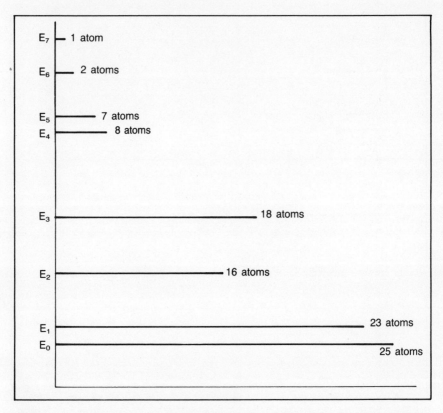

Fig. 7.6 *A population inversion could also be created by bombarding the jar with a monoenergetic electron beam*

the remaining atom is in the ground energy level, E_0. Also suppose a photon comes along whose energy exactly corresponds to the energy difference between the ground level and, say, the second excited state. That is:

$$hc/\lambda = E_2 - E_0$$

where λ is the wavelength of the photon. What happens when the photon interacts with the atom? Because its energy is exactly right, the photon can be absorbed by the atom and the atom will be boosted to the E_2 level. Fig. 7.8a shows before-and-after drawings of the situation.

Now go back and start with the atom already in the E_2 level. What happens this time when the photon interacts with the atom? Because

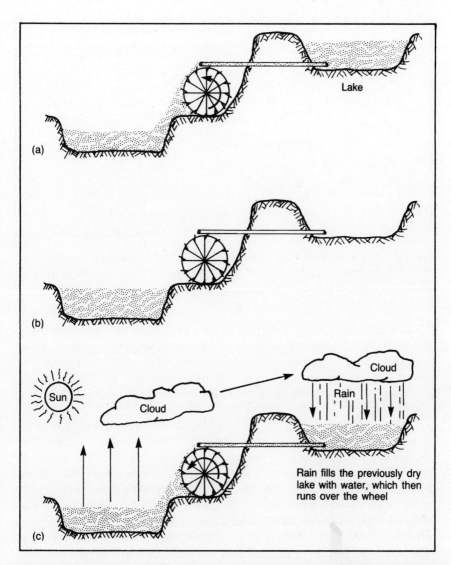

Fig. 7.7 *A system in a nonequilibrium condition (a) and an equilibrium condition (b). Energy from an external source can create a nonequilibrium condition (c)*

its energy is exactly right, the photon can stimulate the excited atom to emit a photon. The atom ends up in the E_0 level, and the two photons depart in the same direction and in phase with each other, as explained in the previous chapter. This sequence of events is shown in Fig. 7.8b.

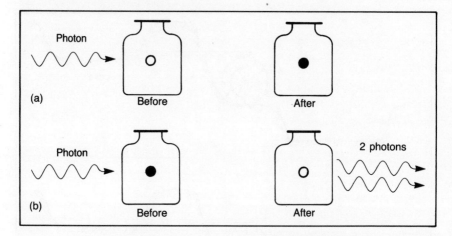

Fig. 7.8 *(a) An atom in the ground state (unshaded) is boosted to an excited state (shaded) when it absorbs a photon; (b) when the atom starts in an excited state, the incident photon can stimulate it to emit*

The next thing to do is to put all 100 atoms back in the jar. Let's say the jar is chilled to absolute zero; all 100 atoms are in the ground state. Suppose three photons come along, each having the correct wavelength, as shown in Fig. 7.9a. The odds are that all three will be absorbed, leaving 97 atoms in the E_0 ground state and 3 atoms in the E_2 excited state.

Next, let's start with 50 of the atoms in the E_2 excited state and 50 in the ground state. (This isn't an equilibrium distribution, so assume that everything happens quickly compared to the time it takes the atoms to revert to their equilibrium distribution.) When the three photons come along, each photon has a 50–50 chance of interacting with an excited atom and a 50–50 chance of interacting with a ground-state atom. If the photon interacts with an excited atom, it will stimulate the atom to emit; if it interacts with a ground-state atom, the photon will be absorbed. So for every photon that's created by stimulated emission, another photon disappears by absorption. The number of photons departing the jar in the "after" picture of Fig. 7.9b is equal to the number of incident photons in the "before" picture. There's no way of telling one photon from another of the same frequency and polarization, but if there were you'd find that the three photons leaving the jar weren't necessarily the same three that entered.

Finally, let's start with all the atoms in the E_2 excited state. Now when the three photons come along, the odds are that every one of them will stimulate an atom to emit, and the number of photons leaving the jar (maybe 6) will be greater than the number that entered.

That's all there is to it.

That's Light Amplification by Stimulated Emission of Radiation. The light is amplified—three in, six out—when stimulated emission adds photons (radiation) to what's already there. It doesn't work without a population inversion, as you can see in Fig. 7.9b. Of course, the 100% population inversion of Fig. 7.9c isn't required. Any population inversion—even 51 excited atoms in the jar—will provide some amplification. But the bigger the population inversion, the bigger the amplification.

Fig. 7.9c is based on the somewhat simplistic assumption that each photon interacts with only a single atom. But if the atoms are big enough in cross section, the photons will interact with atoms soon after they enter the jar and the stimulated photons produced by those inter-

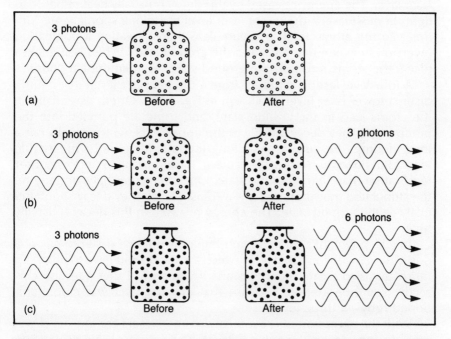

Fig. 7.9 (a) Three photons are absorbed, exciting three atoms from their ground states; (b) if half of the atoms are excited initially, each photon will have a 50–50 chance of being absorbed or of stimulating emission; (c) if all the atoms are excited initially, each incident photon probably will stimulate emission from one of the atoms

actions could stimulate other atoms to emit. (On the other hand, if the atoms are small enough in cross section, one or more of the photons might pass through the jar without interacting with any of the atoms.) So there may or may not be exactly six photons emerging from the jar. The point is that more emerge than enter.

Three-level and four-level lasers

In the example in the previous section, we investigated a population inversion between the ground level and the second excited state. In real lasers usually three or four energy levels are involved in the process of creating a population inversion and then lasing.

In a three-level system, shown in Fig. 7.10a, essentially all the atoms start in the ground state. An external energy source excites them to a *pump level* from which they spontaneously decay quickly to the *upper laser level*. The energy released by this decay is usually heat rather than light. In most lasers, the upper laser level has a long spontaneous lifetime, so the atoms tend to accumulate there, creating a population inversion between that level and the ground state. When lasing takes place, the atoms return to the ground state, each emitting a photon.

A four-level laser is different from a three-level laser in that it has a distinct lower laser level, as shown in Fig. 7.10b. Often, essentially all the atoms start in the ground state, and some are pumped into the pump level. They decay quickly to the upper level, which usually has a long lifetime. (Because it has a long lifetime, it's called a *metastable* level.) But now when lasing takes place, the atoms fall to the lower laser level rather than to the ground state. Once the atoms have undergone the stimulated transition to the lower laser level, they decay spontaneously to the ground state. The energy released in this decay is usually heat.

In which type of system do you think it would be easier to create a population inversion? That is, in which system do you have to pump up more atoms to create the inversion? If you answer that question correctly, you'll be amused to learn that the first laser was chromium-doped ruby—a three-level laser.

With the exception of ruby, most common lasers are four-level systems. In some lasers the pump level isn't a single level; it's a collection of several levels that would be more correctly designated as a *pump band*. And in some common lasers—the helium-neon laser, for example—the pump band is in a different atom than the laser levels. Nonetheless, all these lasers function essentially like the simplified model of Fig. 7.10b.

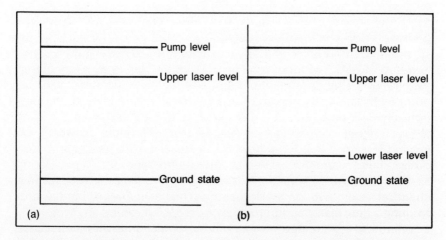

Fig. 7.10   *An energy-level diagram for a three-level laser system (a) and a four-level laser system (b)*

Pumping mechanisms

How can the energy be input to a collection of atoms or molecules to create a population inversion? You know the energy can't be put in thermally; that would just heat the collection, not create a population inversion. Earlier in the chapter, we explored two approaches. The population inversion in Fig. 7.5 was created by plucking some atoms out of the bottom energy level so that it had less population than the level above it. The very first laser-like device ever built, the ammonia *maser* (Microwave Amplification by Stimulated Emission of Radiation) created a population inversion that way. Ammonia molecules were forced through a filter that selectively blocked molecules in the lower of two energy levels so the molecules that emerged exhibited a population inversion. But that technique has not proven successful with modern lasers.

However, the second technique, electrical pumping, is practical for lasers. If the laser medium is placed in an electron beam, the electrons can create a population inversion by transferring their energy to the atoms when they collide. Several types of high-power gas lasers are pumped this way. More common is another technique of electrical pumping, the direct discharge. In this case an electric discharge is created in a tube containing the gaseous laser medium, similar to the discharge in a fluorescent lamp. A population inversion is created in the ions or atoms of the discharge when they absorb energy from the current. Helium-neon lasers and most other common gas lasers are pumped by an electrical discharge.

In other common lasers, like chromium-doped ruby and neodymium-doped YAG, the atoms that lase are embedded in a solid material instead of being in gaseous form. These lasers cannot easily be pumped by an electrical current or an electron beam. Instead, they are optically pumped. The laser material is bombarded with photons whose energy corresponds to the energy difference between the ground level and the pump levels. The atoms absorb energy from the pump photons and are excited to their pump bands.

Chemical energy sometimes can create a population inversion. In a chemical laser, two or more materials react, liberating energy and forming a new material. But the new material—it can be an element or a compound—is created with a built-in population inversion because many of its atoms or molecules have been excited by the chemical energy. Chemical lasers are not very common; but because they're capable of very high powers, they're of special interest for one particular application: laser weapons for the military.

There is also an exotic pumping mechanism mentioned here only for the sake of completeness. Lasers have been pumped by nuclear particles, usually from nuclear bombs. The energy in the particles creates a population inversion in the laser medium, and a pulse of laser output is obtained before the shock wave and other energy from the bomb destroys the laser. Such lasers may someday have military application.

Finally, free-electron lasers are pumped by high-energy electrons. The output of a free-electron laser is produced from electrons that are free rather than bound to an atom or molecule. So the physics of a free-electron laser is much different than other lasers. Free electrons have no fixed energy levels like bound electrons, so the laser's wavelength can be tuned. Although free-electron lasers are promising for future applications, they are highly experimental and have been operated in only a few laboratories worldwide.

Questions

1. The mathematical expression of Boltzmann's Law is:

$$N_i = N_0 \exp - (E_i/k_B T)$$

where N_i is the number of atoms (or molecules) in the ith level whose energy above ground state is E_i and N_0 is the number of atoms in the ground state. Boltzmann's constant is 1.38×10^{-23} J/K, and T is the temperature.

Use this form of Boltzmann's Law to calculate the number of atoms in the upper laser level of a chromium-doped ruby laser rod that contains 2×10^{19} chromium ions. Assume the rod is at room temperature (T = 300K). (Hint: First calculate $E_i = hc/\lambda$. The wavelength of a ruby laser is 694.3 nm.)

2. Why it is impossible to have a continuouswave optically pumped two-level laser?

3. Name several other examples on systems not in thermal equilibrium. For each of your examples, what is the approximate "spontaneous lifetime," i.e., the time it takes the system to relax into thermal equilibrium?

Eight

Laser Resonators

This chapter will introduce the concept of the laser resonator, the crucial device that provides the feedback necessary to make a laser work. In Chapter 7 we found that there must be a population inversion for stimulated emission to occur. In this chapter we'll find that the resonator is necessary if the stimulated emission is going to produce a significant amount of laser light.

It turns out that a resonator isn't absolutely necessary to make a laser work, but as you'll see in the first section of this chapter, a laser without a resonator usually just isn't very practical.

Then we'll discuss the concept of circulating power, the light that literally circulates back and forth between the mirrors of a laser resonator. We'll see how that power experiences both gain and loss as it circulates, and we'll come to the important conclusion that, in a steady-state laser, the gain must equal the loss. We'll examine the concept of gain saturation that allows a laser to satisfy this requirement.

Finally, we'll discuss an important class of resonators called *unstable resonators* that can sometimes produce more output than the more conventional stable resonators.

Incidentally, there are two terms you'll come across that are used interchangeably in laser technology: *resonator* and *cavity*. The second word is a holdover from microwave technology because a microwave oscillator is a completely enclosed cavity. Because the word "cavity" can mean several things in laser technology, we will not use the word to mean "resonator."

Why a resonator?

Let's think about what kind of a laser you might have with just a population inversion and nothing else—specifically, no resonator. Suppose you created a big population inversion in a laser rod, and suppose that one of the atoms at the far end of the rod spontaneously emitted a photon along the axis of the rod. The photon might stimulate another atom to emit a second photon and, since we're assuming a big population inversion, one of these two photons might even find a third excited atom and create another stimulated photon. So, screaming out the near end of the laser rod come . . three photons.

Now, the energy in a visible photon is about 10^{-19} J, while a real laser might produce 1 J. Obviously, we have a way to go from the three photons we've generated so far.

What next?

Well, we might try adding more laser rods, as shown in Fig. 8.1. If each incoming photon in the second rod results in three output photons, we'll have 9 photons coming out of the second rod. A third rod will result in 27 photons.

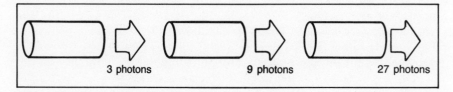

3 photons   9 photons   27 photons

Fig. 8.1 *How to make a laser without a resonator*

But we want roughly 10^{19} photons to produce 1 J of output, and it should be obvious that it's going to take a long time to get there. Is there a better way?

The better way, of course, is to use mirrors, as shown in Fig. 8.2. The photons are reflected back and forth for many passes through the rod, stimulating more and more emission on each pass. As indicated in the figure, the mirrors can be gently curved so they tend to keep the light concentrated inside the rod. One of the mirrors is 100% reflective, but

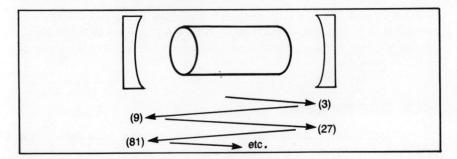

Fig. 8.2 *How to make a laser with a resonator*

the other mirror transmits part of the light hitting it. This transmitted light is the output beam from the laser. The transmission of the output mirror varies from one type of laser to another but is usually somewhere between 1% and 50%.

For the sake of completeness, we should mention here that some lasers have such enormous gain that 10^{19} photons can be produced in a single pass. These *super-radiant* lasers do not need resonators. Nitrogen lasers sometimes operate in this fashion.

Circulating power

If the photons in Fig. 8.2 bounce back and forth between the mirrors for a long enough period, the laser will reach a steady-state condition and a relatively constant power will circulate between the mirrors. This *circulating power* is not absolutely constant, as indicated in Fig. 8.3. Part of it is lost when it hits the output mirror, and this lost power is replaced when the light passes through the gain medium.

Fig. 8.3 shows how the circulating power varies inside the resonator. You can follow it around, starting, for example, at point A and moving to the right. The circulating power drops at the output mirrors because part of the light is transmitted through the mirror. The remaining light travels through the gain medium where it is partially replenished. There's a small loss at the back mirror because no mirror is a perfect reflector. Then the light returns for a second pass through the gain medium, where it is fully restored to its previous level at point A.

Have you realized, as you've read this, that the power in the output beam is determined by the amount of circulating power and the transmission of the output mirror? Suppose that there are 50 watts of circu-

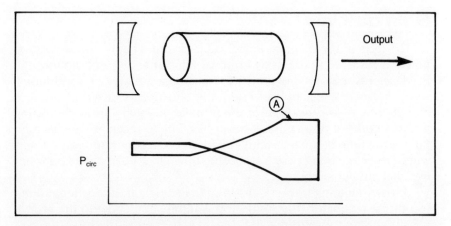

Fig. 8.3 *A schematic representation of the circulating power inside a laser resonator; the lower figure shows the power in the resonator*

lating power inside the laser and the output mirror has 2% transmission. Then the output beam is simply 2% of the circulating power that is incident on the mirror, or 1 watt. Mathematically, the concept is expressed this way:

$$P_{out} = \tau P_{circ}$$

where τ is the transmission of the output mirror and P_{out} and P_{circ} are the output and circulating powers, respectively.

You can also use this concept to calculate the circulating power inside a resonator. While a direct measurement of the circulating power would be difficult to make, it's easy to measure the output power and then simply divide by the mirror transmission to find what the circulating power really is.

The equation above says that the output power is proportional to the mirror transmission. So does that mean you can always increase the output power by increasing the mirror transmission? No, because as the transmission of the output mirror is increased, the circulating power will decrease. Whatever happens to the output power depends on whether the circulating power decreases faster than the mirror transmission increases, and this depends on the particular laser you're looking at. In fact, for any laser there will be an optimum value for the transmission of the output mirror that will produce the maximum possible output power.

Gain and loss

The loop in the diagram in Fig. 8.3 is closed. That is, the circulating power is restored to precisely its initial value after a round trip through the laser resonator. This is true for any steady-state, or continuous-wave, laser. In most pulsed lasers the situation is different because energy moves so quickly from the population inversion to circulating power to output power that it never has time to reach an equilibrium. But for the time being let's limit our attention to continuouswave lasers where the circulating power can settle down to a steady-state behavior like that shown in Fig. 8.3.

If the circulating power is restored to its original value after a round trip of the resonator, the round-trip gain must be equal to the round-trip loss. If the round-trip gain is less than the round-trip loss, the laser will not lase. On the other hand, if the round-trip gain is greater than the round-trip loss, the gain will *saturate* until it is reduced to the same value as the round-trip loss. We'll discuss this saturation phenomenon in more detail later in this section.

It's important to take note of the fact that we have just identified a second requirement for lasing. In Chapter 7 we found that there must be a population inversion in the gain medium. In this chapter we see that merely having a population inversion isn't enough; the population inversion must be large enough so the round-trip gain is at least as large as the round-trip loss. In the parlance of laser technology, the gain that is just barely sufficient for lasing is called the *threshold gain*.

What are the causes of loss inside a laser resonator? Obviously, the transmission of the output mirror is one source of loss, but some circulating power is also lost at every optical surface in the resonator because there's no such thing as a perfect surface. Some light will be scattered and reflected no matter how well the surface is polished. As we've said, some light is lost at the imperfect rear mirror. Other light is lost by scattering as it propagates through refractive-index inhomogeneities in the gain medium. Some loss by diffraction occurs because of the finite aperture of the laser beam inside the resonator. Altogether, these resonator losses add up to a few percent per round trip in most continuouswave resonators.

Now, if you think about it for a second, it seems as if we've gotten ourselves into a contradictory situation. The round-trip loss depends solely on the passive qualities of the resonator—its mirror transmission, etc.—and the round-trip gain must be equal to this round-trip loss. Do you see the problem? The problem is that we determined what the round-trip gain is without knowing anything about how hard the laser is being pumped. That is, we're apparently saying that the gain is independent of how hard the laser is pumped. Can that be right?

It turns out that it is. To understand why, you must understand that there are two kinds of laser gain: saturated gain and unsaturated gain. The difference between the two is diagramed in Fig. 8.4. Unsaturated gain is sometimes referred to as small-signal gain because it's the gain observed with only a very small input signal. In Fig. 8.4a, 100 photons are amplified to 102 photons by the laser rod. The unsaturated gain is therefore 2%. But in Fig. 8.4b the same laser rod provides a gain of only 1% when 10^{20} photons pass through it. What happened? So many photons passed through the rod, stimulating emission as they went, that the population inversion was significantly depleted by stimulated emission. So the gain was reduced—saturated—from its unsaturated value in Fig. 8.4a.

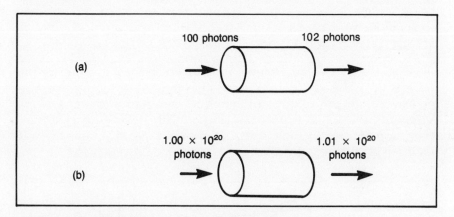

Fig. 8.4 *Unsaturated gain (a) and saturated gain (b)*

In the unsaturated case of Fig. 8.4a, atoms pumped to the excited state are not stimulated to emit. These atoms can get out of the excited state only by spontaneous emission or by collisional de-excitation processes.[1] But in the saturated case of Fig. 8.4b, about 10^{18} atoms leave the excited state by stimulated emission. The absence of these 10^{18} atoms in the excited state accounts for the gain difference between the two cases.

---

[1]*Two atoms are stimulated to emit the two extra photons produced in Fig. 8.4a. But the remaining 10^{19} atoms must get out of the excited state some other way.*

Now how does this resolve the seeming contradiction from several paragraphs above? Let's start by thinking about a continuouswave laser sitting on a table, lasing away with a 3% round-trip loss and a 3% round-trip gain. What happens when you turn up the pump power? Well, instantaneously the population inversion gets bigger and so does the gain. But this larger gain produces more circulating power, and the increased circulating power causes the gain to saturate more than it was before. The gain quickly decreases back to 3%, but the circulating power is now greater than it was before you turned up the gain. On the other hand, if you turn down the pump power, the reverse process takes place and you wind up with less circulating power but the same 3% gain.

So you see that the actual gain in a laser resonator is in fact independent of the pump power. It's the circulating power, and hence the output power, that varies with the pump power.

Unstable resonators

Most lasers have stable resonators in which the curvatures of the mirrors keep the light concentrated near the axis of the resonator. If you trace the path of a ray of light between the mirrors of a stable resonator, you'll find that the ray is eventually reflected back toward the resonator axis by the mirrors, as shown in Fig. 8.5. The only way light can escape from the resonator is to go *through* one of the mirrors.

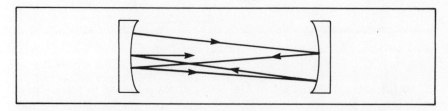

Fig. 8.5 *A ray is always reflected back toward the center by the curved mirrors of a stable resonator*

In the unstable resonator diagramed in Fig. 8.6, the light rays keep on moving away from the resonator axis until eventually they miss the small convex mirror altogether. The output beam from this resonator will have a doughnut-like shape with a hole in the middle caused by the shadow of the small mirror. (There are clever ways to design unstable resonators that avoid the hole-in-the-center beam.) The advantage of unstable resonators is that they usually produce a larger beam volume

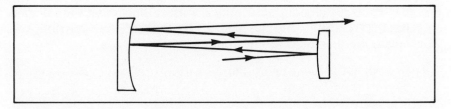

Fig. 8.6 *In this unstable resonator, a ray will eventually be reflected past one of the mirrors*

inside the gain medium so the beam can interact with more of the population inversion and thereby produce more output power. Unstable resonators are usually used only with high-power, pulsed gas and solid-state lasers.

Incidentally, don't be confused about the use of the word "stable" here. A stable resonator is, by definition, one in which a ray is trapped between the mirrors by their curvature. The word stable implies nothing about a resonator's sensitivity to misalignment nor about an absence of fluctuations in its output power.

Questions

1. Suppose you wanted to make a laser that could produce 1 J (10^{19} photons) without mirrors, as indicated in Fig. 8.1. How many laser rods would you need?

2. Calculate the circulating power inside a continuouswave Nd:YAG laser that produces a 10-W output beam and has a 6% transmissive output mirror. Calculate the circulating power inside a HeNe resonator that has a 1% mirror and produces 2 mW.

3. Consider the laser shown in the diagram below. If the output mirror

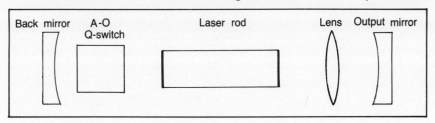

Back mirror   A-O   Laser rod   Lens Output mirror
                Q-switch

has 3% transmission, each optical surface (including the mirrors) has 0.25% scatter, and the back mirror has 0.5% transmission, calculate the threshold round-trip gain.

4. In a Nd:YAG laser, the unsaturated gain is related to the saturated gain by the equation:

$$g = \frac{g_0}{1 + \beta P_c}$$

where g is the saturated gain, g_0 is the unsaturated gain, P_c is the circulating power, and β is the saturation parameter, an experimentally determined quantity. From this equation and the equation $P_{out} = \tau P_{circ}$, derive an equation expressing the output power from the laser in terms of unsaturated gain (g_0), saturated gain (g), saturation parameter (β), and mirror transmission (τ).

If the laser in problem 3 has an unsaturated gain of 5% and a saturation parameter of 0.018 W^{-1}, calculate its output power. (Hint: the saturated gain is equal to one-half the round-trip loss. Can you explain why this is true?)

Calculate the output power if the 3% mirror is replaced by a 2% mirror. By a 4% mirror.

Chapter

Nine

Resonator Modes

In Chapter 8 we learned why a resonator is necessary and how the optical power circulates back and forth between the mirrors of a laser resonator. In this chapter we want to take a close look at that circulating power, and in particular we want to examine its spatial distribution within the resonator.

We'll find that studying the *transverse* spatial distribution of energy within the resonator leads to an understanding of how a laserbeam propagates through space outside the resonator. This chapter will introduce some equations that you can use to calculate how the characteristics of a beam change as it propagates. You'll also learn how to select the proper resonator mirrors to produce a laserbeam of a given size and divergence.

But there are some resonator configurations that are inherently unstable, and in this chapter you'll learn how to determine whether a resonator is stable.

Finally, we'll see how the *longitudinal* spatial distribution of energy within a resonator affects the output beam from the laser.

Spatial energy distributions

It is recorded that an early philosopher in the field of science tried to capture light in a Greek vase. Standing outdoors on a sunny afternoon, he turned the open mouth of the vase toward the sun and let light flood into the container. Quickly, he slipped a lid over the mouth of the vase and hurried into a darkened cave, where he carefully removed the lid to let the light escape. But he was disappointed, for the interior of the vase was pitch black every time he performed the experiment.

Of course, you know that the experiment was doomed to failure. But do you really know why? What happened to the light that had been in the vase? Where did it go? And how long did it take to go wherever it went?

At one instant the vase was full of photons, bouncing around off the sides of the vase, and the next instant the vase was capped and all the photons had disappeared. Where did they go? The photons were absorbed by the walls of the vase. Even a polished white Greek vase absorbs maybe 10% of the light incident on it. So after the light has bounced around 100 or so times inside the vase, it's practically all gone. How long does that take? Well, if you figure it's a big Greek vase 1 ft in diameter, and if you remember that the speed of light is roughly 1 ft/ns, you will see that the light will be gone after about 100 ns. So the philosopher actually had the right idea; he just didn't move fast enough.

In the ensuing centuries, modern science has improved on the Greek vase as a light-storing device. A laser resonator is, in fact, nothing more than a modern light storage device. Admittedly, it is designed with a figurative hole in it because the output mirror allows part of the stored energy to "leak" out. But the gain medium replaces the energy as fast as it is lost through mirror transmission and the other losses discussed in Chapter 8.

When you talk about *resonator modes*, you're talking about the spatial distribution of stored light energy between the laser mirrors. It turns out that energy isn't stored uniformly in a resonator, like water in a glass. Instead, the energy exists in clumps, somewhat like cotton balls stored in a jar. The resonator mode is determined by the spatial arrangement of these clumps of light energy.

There are two kinds of modes: *transverse* and *longitudinal*.

To visualize the transverse modes of a laser, imagine that the resonator is cut in half along a plane transverse to the laser axis, as shown in

Fig. 9.1a. If you then examined the distribution of energy along this plane, you'd see the shape of the transverse laser mode. On the other hand, if you sliced the resonator as shown in Figure 9.1b, you'd see the shape of the longitudinal laser mode.

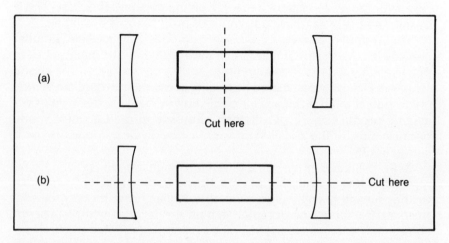

Fig. 9.1 *How to visualize transverse laser modes (a) and longitudinal laser modes (b)*

This concept is analogous to mapping the spatial distribution of cotton balls in a jar, as shown in Figure 9.2. You can imagine drawing a map corresponding to the distribution of cotton along the plane of the figure. But in a laser resonator the energy map doesn't change if the plane moves. That is, if the "cut-here" plane of Fig. 9.1a moves to the left or right (or up or down in Fig. 9.1b), the shape of the energy distribution you'd see in that plane doesn't change (although its size might).

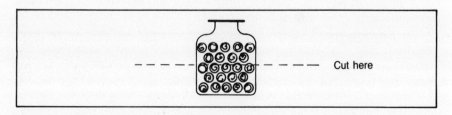

Fig. 9.2 *How to visualize horizontal cotton-ball storage modes*

Transverse resonator modes

It's not necessary to cut the resonator in half, as in Fig. 9.1a, to see the shape of the transverse modes of a resonator. All you need to do is to look at the shape of the output beam because the pattern inside the resonator moves out through the mirror and becomes the shape of the beam. The beam can have a number of profiles, as shown in Fig. 9.3.

Theoretically, dozens of transverse modes can oscillate simultaneously in a resonator and each can have a different frequency, but in practice only several (or sometimes only one) oscillate. The mode shapes in Fig. 9.3 were created by forcing a resonator to oscillate in only one mode at a time. But a laser oscillating in many modes won't necessarily produce more power than one oscillating in a single mode because most of the available power must be divided among the oscillating modes.

Note that each mode has a different designation in Fig. 9.3. If you remember that the number of dark stripes in the pattern corresponds to the subscript, you'll always be able to name the mode properly. Incidentally, there's no accepted system for deciding which subscript comes first; one person's TEM_{41} mode is another's TEM_{14}, and both are correct.[1]

As shown in Fig. 9.4, high-order modes are larger than low-order modes. For many laser applications, it's important that the laser oscillate only in the TEM_{00} mode. How can you prevent a laser from oscillating in its higher-order modes?

The answer has to do with the relative sizes of the different modes. The TEM_{00} mode is smaller in diameter than any other transverse mode. So if you place an aperture of the proper size—as shown in Fig. 9.4—inside the resonator, only the TEM_{00} mode will fit through it. Higher-order modes will be extinguished because the loss imposed on them by the aperture will be greater than the gain provided by the active medium. Some TEM_{00} lasers come equipped with apertures like the one shown in Fig. 9.4, while in others the small diameter of the active medium acts as an effective aperture.

In Fig. 9.4 the TEM_{11} mode occupies a larger volume in the gain medium than the TEM_{00} mode does. The TEM_{11} mode can therefore interact with more of the population inversion and extract more power from the laser. For this reason lasers oscillating in high-order modes

---

[1]*TEM stands for "transverse electromagnetic," a name that derives from the way the electric and magnetic fields behave at the resonator's boundary.*

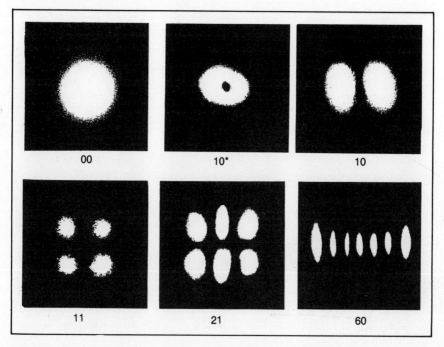

Fig. 9.3 *The shapes of transverse laser modes*

usually produce more power than otherwise similar lasers limited to TEM_{00} oscillation. But the advantages of the TEM_{00} mode very often outweigh the cost of reduced power.

Gaussian-beam propagation

The TEM_{00} mode is so important that there are several names for it in laser technology, all meaning the same thing. The TEM_{00} mode is called the "Gaussian mode," the "fundamental mode," or even the "diffraction-limited mode." No matter what it's called, it's a very important mode, and this section will describe how the light produced by a Gaussian-mode laser propagates through space.

But before you can understand how a Gaussian beam propagates, you must understand two parameters that characterize the beam. One is easy to understand; the other is a little more subtle. The easy one is the *beam radius,* the radius of the spot the beam would produce on a screen. It's a somewhat arbitrary parameter because a Gaussian beam

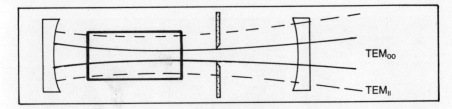

Fig. 9.4   *An aperture in the resonator can force it to oscillate only in the TEM$_{\infty}$ mode*

doesn't have sharp edges. The intensity profile of a Gaussian beam is given by the equation:

$$I = I_oe^{-2x^2/w^2}$$

where I_o is the intensity at the center, x is the distance from the center, and w is the beam radius. The intensity profile is pictured in Fig. 9.5. The "edge" of this beam is defined to be the point where its intensity is down to $1/e^2$ (about 13%) of its intensity at the center. To the eye, the place where the intensity has dimmed to $1/e^2$ of its maximum value looks like the edge of the beam.

The other parameter that characterizes a Gaussian beam—the more subtle parameter—is the *radius of curvature* of the beam's wavefront. Remember that one of the characteristics of coherent light is that all the waves are in phase with each other. If you were to construct a surface that intersected all the points of common phase in a Gaussian beam, that surface would be spherical. The idea is shown graphically in Fig. 9.6. The dotted spherical surface in this figure passes through the trough of each wave in the beam.

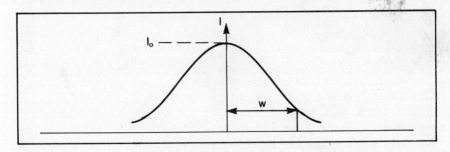

Fig. 9.5   *The intensity profile of a Gaussian beam*

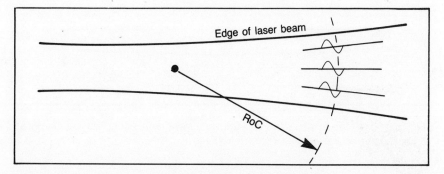

Fig. 9.6 *In a Gaussian beam, the surface of constant phase (dotted line) is spherical*

Both the beam radius and the radius of curvature change as the beam propagates. You can think of a Gaussian beam having the appearance shown in Fig. 9.7. (Of course, you can't really see the radii of curvature in the beam, but you can think about their being there.) The light could be traveling either direction in this drawing—left to right or right to left. The beam radii and the radii of curvature would be the same in either case. The radius of curvature is infinite at the beam waist, drops sharply as you move away from the waist, and then begins increasing again as you move further from the waist. At long distances from the waist, the radius of curvature is equal to the distance from the waist. The beam radius increases steadily, of course, as you move away from the waist.

The equations below allow you to calculate both the beam radius (w) and the radius of curvature (R) at any distance (z) from the waist if you know the beam radius at the waist (w_o) and the laser wavelength (λ).

$$w = w_o \left[1 + \left(\frac{\lambda z}{\pi w_o^2} \right)^2 \right]^{\frac{1}{2}}$$

$$R = z \left[1 + \left(\frac{\pi w_o^2}{\lambda z} \right)^2 \right]$$

The first equation gives the beam radius, and the second gives the wavefront radius of curvature. Let's look at some examples where these equations would be useful.

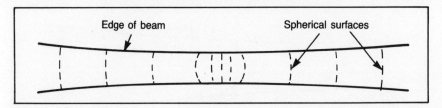

Edge of beam   Spherical surfaces

Fig. 9.7 *Both the beam radius and the wavefront radius of curvature change as a Gaussian beam propagates through space*

Suppose you wanted to do a laser-ranging experiment to measure the distance from the earth to the moon. You'd aim a pulse of laser light at the retroreflector left on the moon by the astronauts. By very carefully measuring how long it took the pulse to travel to the moon and back, you'd be able to figure out the distance with an accuracy of feet or even inches. But how wide is the beam by the time it gets to the moon? If it's too wide, only a tiny fraction of the light will be reflected back by the retroreflector, and the amount that returns to earth could be too small to detect.

Let's suppose that we're using a Nd:YAG laser whose wavelength is 1.06 μm and whose beam waist is 0.5-mm radius. The approximate earth-moon distance is 239,000 miles. We must invoke the first of the two equations:

$$w = w_o \left[1 + \left(\frac{\lambda z}{\pi\, W_o{}^2} \right)^2 \right]^{1/2}$$

Let's write down the known quantities and express them all in the same dimension, meters:

$$\lambda = 1.06 \times 10^{-6}\ \text{m}$$
$$w_o = 5 \times 10^{-4}\ \text{m}$$
$$z = 239,000\ \text{mi} = 3.84 \times 10^8\ \text{m}$$

Next, substitute the known values into the equation:

$$w = (5 \times 10^{-4}\ \text{m}) \left[1 + \left(\frac{(1.06 \times 10^{-6}\ \text{m})\ (3.84 \times 10^8\ \text{m})}{\pi\ (5 \times 10^{-4}\ \text{m})^2} \right)^2 \right]^{1/2}$$

Finally, do the arithmetic:

$$W = (5 \times 10^{-4} \text{ m}) [1 + (5.2 \times 10^8)^2]^{1/2}$$
$$\approx (5 \times 10^{-4} \text{ m}) (5.2 \times 10^8) = 2.5 \times 10^5 \text{ m}$$

This beam is too wide. The retroreflector is only about a meter in diameter, so only a minuscule fraction of the light will be reflected back toward earth. What could you do to decrease the size of the beam on the moon?

Take a look at the equation again. There are three parameters: λ, z, and w_0. The laser wavelength is fixed, and you can't move the moon any closer to the earth to make the problem easier. But you can adjust the radius of the laserbeam, and that's the solution. The larger the waist of a Gaussian beam, the smaller its divergence will be. If you expand the beam with a telescope before it leaves earth, you can greatly reduce its divergence.

There are many different Gaussian beams, one for each size of waist. If you know the size of the beam at its waist, you can calculate the beam size and its radius of curvature at any point in space. The characteristics of a Gaussian beam are completely determined from its waist size (assuming, of course, that the wavelength does not change). By using a lens—or a system of lenses or focusing mirrors—it is possible to convert one Gaussian beam into another. In Fig. 9.8 a small-waisted divergent Gaussian beam is changed into a less-divergent, larger-waisted beam by a lens.

You can imagine numerous applications for the first of the two equations given above. A manufacturer making laser surveying equipment would want to know how big the beam would be a few hundred yards from the laser so he could make his detectors the proper size. An engineer designing a laser scanner for a grocery store would need to know the size of the beam on the window where the groceries sweep

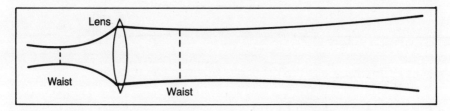

Fig. 9.8 *A lens converts one Gaussian beam into another*

across. But where would you use the second equation? What good does it do to know the radius of curvature of a Gaussian beam?

For a Gaussian beam to exist in a resonator, its wavefronts must fit exactly into the curvature of the mirrors. The Gaussian beam from Fig. 9.7 is shown again in Fig. 9.9, together with some of the resonators that would support this particular beam. For example, the portion of the beam between E and H could oscillate in a resonator composed of a flat mirror and a mirror whose curvature matched the wavefront at H. Or two curved mirrors could match the beam's wavefronts at B and H. It's even possible to have a stable resonator configuration with a convex mirror, as shown in the bottom drawing of Fig. 9.9.

It is easy to understand why the wavefront of a Gaussian beam must fit exactly into the curvature of a resonator mirror. In a Gaussian beam (or any wave) energy flows perpendicular to the wavefront. If you want to see the direction of energy flow in a wave, simply visualize little arrows all along the wavefront, perpendicular to the wavefront where they meet it. If the mirror curvature exactly fits the wavefront, all the energy in the wave is exactly reflected back on itself and the resonator is stable.

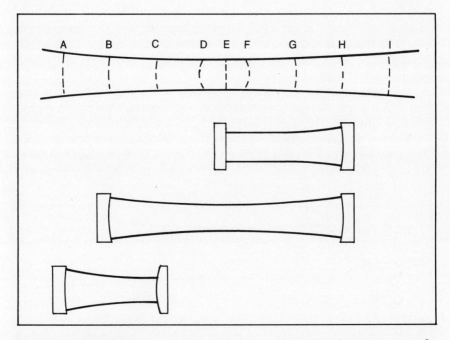

Fig. 9.9 *A resonator supports a Gaussian beam whose wavefront curvatures fit the mirror curvatures*

Now do you see the value of the second equation? It tells you what mirrors you must use to produce a given Gaussian beam in a resonator. Let's look at an example. Suppose you wanted to design an argon laser (λ = 514.5 nm) whose beam had a 0.5-mm diameter right at the center of the laser and whose mirrors were 1 m apart. What mirrors would you use?

You would want to calculate the wavefront radius of curvature at the points where the mirrors are to go, then obtain mirrors with the same curvature. Begin by writing the equation you'll use:

$$R = z \left[1 + \left(\frac{\pi \, w_o^2}{\lambda z} \right)^2 \right]$$

Then write down the known parameters, again putting everything in meters:

$$Z = 5 \times 10^{-1} \text{ m}$$
$$\lambda = 5.14 \times 10^{-7} \text{ m}$$
$$w_o = 2.5 \times 10^{-4} \text{ m}$$

Substitute the values into the equation:

$$R = (5 \times 10^{-1} \text{ m}) \left\{ 1 + \left[\frac{\pi \, (2.5 \times 10^{-4} \text{ m})^2}{(5.14 \times 10^{-7} \text{ m}) \, (5 \times 10^{-1} \text{ m})} \right]^2 \right\}$$

Finally, do the arithmetic:

$$R = (5 \times 10^{-1} \text{ m}) \, [1 + (0.76)^2]$$
$$= (5 \times 10^{-1} \text{ m}) \, (1.58)$$
$$= 0.79 \text{ m, or } 79 \text{ cm}$$

So you'd want to use mirrors whose radius of curvature was about 80 cm.

What about the physical size of the mirrors? Of course, they must be larger than the diameter of the beam when it hits the mirrors. Could you calculate that size?

A stability criterion

As we learned in Chapter 8, a stable resonator is one in which rays can be trapped by the curvature of the mirrors—they will bounce back and forth between the mirrors forever. Now that we know about

Gaussian beams, a second definition is possible: A stable resonator is one for which a Gaussian beam can be found whose wavefronts fit the curvatures of the mirrors. Obviously, some possible configurations are excluded, such as the one having two convex mirrors. But in general it's difficult to tell just from casual inspection whether a resonator is stable. Fig. 8.6 showed a resonator with concave and convex mirrors that is unstable, and Fig. 9.9 shows a similar-looking concave/convex resonator that is stable. Apparently, the difference depends on the exact curvatures of the mirrors. What's more, even a resonator that has two concave mirrors isn't necessarily stable. In this case as well, stability depends on the exact curvature of the mirrors. How can you tell? Is there some test you can perform to find out if a particular configuration is a stable resonator?

Fortunately, there is. Otherwise, the only way to know for sure would be to construct the resonator and try to make it lase—a difficult and tedious task. Fig. 9.10 shows the parameters you need to calculate the stability of a resonator. The curvatures of the two mirrors are r_1 and r_2, and the spacing between them is ℓ. If the mirror is convex, then its radius is taken to be negative. The condition for stability is:

$$0 \leq g_1 g_2 \leq 1$$

where g_1 and g_2 are the so-called "g-parameters" defined in Fig. 9.10. If their product is between zero and one, the resonator is stable. Let's look at some examples.

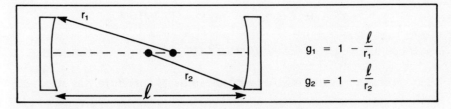

$$g_1 = 1 - \frac{\ell}{r_1}$$

$$g_2 = 1 - \frac{\ell}{r_2}$$

Fig. 9.10 *The g-parameters for calculating resonator stability*

Fig. 9.11 shows a concave/convex configuration. Is it stable? First, calculate the g-parameters:

$$g_1 = 1 - \frac{50 \text{ cm}}{-500 \text{ cm}} = 1.1 \qquad g_2 = 1 - \frac{50 \text{ cm}}{100 \text{ cm}} = 0.5$$

Then multiply them together:

$$g_1g_2 = (1.1)\ (0.5) = 0.55$$

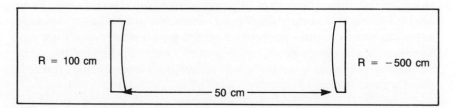

Fig. 9.11 *A stable concave-convex resonator configuration*

Because 0.55 is between one and zero, this particular configuration is stable and will support a Gaussian mode or any of the higher-order modes shown in Fig. 9.3. It is interesting to note that stability depends only on the mirror curvature and separation, not on laser gain, laser wavelength, or any other characteristic. The resonator of Fig. 9.11 is stable for a HeNe laser, a Nd:YAG laser, or a carbon dioxide laser.

Fig. 9.12 shows a concave/concave mirror configuration. To find if it's stable, first calculate the g-parameters:

$$g_1 = g_2 = 1 - \tfrac{9}{3} = -2$$

Multiply them together:

$$g_1g_2 = (-2)^2 = 4$$

The product is greater than one, so this configuration is not stable.

It's important to understand that a resonator must not only be stable but that its mirrors must be exactly aligned with each other before it can

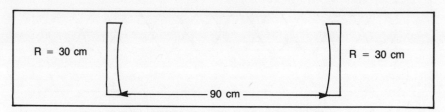

Fig. 9.12 *An unstable concave-concave configuration*

support laser oscillation. A ray cannot be trapped between the mirrors of any resonator, stable or unstable, if those mirrors are misaligned.

It's also important to understand that the stability of a resonator, as defined here, has nothing to do with how sensitive to misalignment the resonator is. Some resonators can have mirrors tilted by relatively large angles before their output power decreases, and others are completely extinguished by even a small tilt. And the g-parameters don't tell you anything about sensitivity to misalignment; indeed, some of the resonators that are most insensitive to misalignment have the g_1g_2 product exactly equal to zero.

Longitudinal modes

The energy stored in a laser resonator has spatial variations not only perpendicular to the laser axis—as shown in Fig. 9.1a—but along the axis as well, as shown in Fig. 9.1b. But these longitudinal variations are much smaller in scale than the transverse variations. Each longitudinal resonator mode is a *standing wave* of light, created by the overlap of two traveling waves that are moving in opposite directions. The spatial distribution of energy in one such longitudinal mode is shown in Fig. 9.13.

In Fig. 9.13, the wavelength of light is necessarily shown far out of proportion to the size of the mirrors. We can calculate how many wavelengths there are between the mirrors of a real laser. Suppose its mirrors are separated by 30 cm and the laser wavelength is one μm. Then the number of wavelengths between the mirrors is:

$$N \times \ell/\lambda = 3 \times 10^{-1} \text{ m}/(1 \times 10^{-6} \text{ m}) = 300,000$$

Fig. 9.13 shows only five wavelengths between the mirrors.

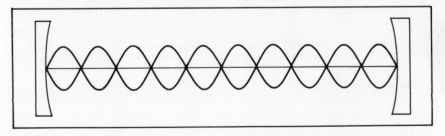

Fig. 9.13 *Energy distribution in a single longitudinal laser mode*

Fig. 9.13 also shows the so-called "boundary condition" for a stand-ing light wave in a resonator: there must be a node at both mirrors. (The node of a wave is the point where the wave passes through zero.) So we are led to an important conclusion: Not just any wavelength will work in a resonator. For a wave to work in a given resonator, there must be room for exactly an integral number of half-wavelengths between the mirrors. You can have 600,000 half-wavelengths between the mir-rors of the laser in the previous paragraph, or you can have 600,001. But you cannot have anything in between.

If you think about it for a second, you will see that there is only a very small wavelength difference between a wave with 600,000 half-wavelengths between the mirrors and one with 600,001. That is, you don't have to squeeze each of the 300,000 wavelengths very much to make room between the mirrors for one more half-wavelength. In fact, the difference turns out to be so small that more than one longitudinal mode can oscillate in a laser at the same time. A laser is not perfectly monochromatic, and the amount of imperfection is usually greater than the wavelength difference between longitudinal modes.

Let's calculate the frequency spacing between adjacent longitudinal modes of a laser resonator. In the calculation we'll use some of the concepts about wavelength and frequency introduced in Chapter 2. Begin with the requirement on wavelength: there must be an integral number of half-wavelengths between the mirrors. This can be ex-pressed mathematically by the equation:

$$n \frac{\lambda}{2} = \ell$$

where n is the integer and ℓ is the mirror spacing. Solve for wave-length:

$$\lambda = \frac{2\ell}{n}$$

From Chapter 2, $f = c/\lambda$, so:

$$f_n = n \frac{c}{2\ell}$$

where c is the speed of light. The frequency of the next mode—the one with n + 1 half-wavelengths between the mirrors—is:

$$f_{n+1} = (n + 1) \frac{c}{2\ell}$$

The difference between these two frequencies is:

$$\Delta f \equiv f_{n+1} - f_n = \frac{c}{2\ell}$$

Does this equation look familiar? It's the equation for the resonant frequencies of a Fabry-Perot interferometer, discussed in Chapter 4. Now you know where those frequencies came from: each one is a different standing wave between the mirrors of the interferometer.

It's interesting that the frequency spacing depends only on the spacing between the resonator mirrors and not on the laser wavelength. So the frequency spacing between longitudinal modes of a 30-cm HeNe resonator is the same as the frequency spacing between longitudinal modes of a 30-cm Nd:YAG resonator.

Let's calculate what that spacing is:

$$\Delta f = \frac{c}{2\ell} = \frac{3 \times 10^8 \text{ m/s}}{2 \, (3 \times 10^{-1} \text{ m})} = 5 \times 10^8 \text{m/s}$$

The bandwidth of a typical HeNe or Nd:YAG laser is many times larger than 500 MHz, so many longitudinal modes can oscillate at the same time in the resonator. Fig. 9.14 shows what two simultaneously oscillating longitudinal modes might look like. Fig. 9.15 shows what the frequency spectrum of the longitudinal modes looks like. Modes a and

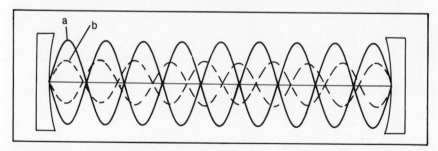

Fig. 9.14 *Two longitudinal laser modes*

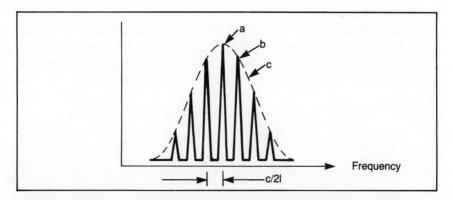

Fig. 9.15   *The frequency spectrum of longitudinal laser modes*

b of this figure correspond to the two modes in Fig. 9.14. (Not all the modes of Fig. 9.15 are shown in Fig. 9.14.)

The shape of the dotted curve in Fig. 9.15 is determined by the gain of the active medium, as we'll discuss in more detail in the next chapter. There is no mode at *c* in Fig. 9.15 because an integral number of wavelengths of light at that frequency would not fit between the mirrors.

Questions

1. Suppose a Cr:Ruby laser (λ = 694.3 nm) is used to track an earth satellite. If the beam has a waist diameter of 2 mm at the laser and the satellite is 500 miles straight up, calculate the diameter of the beam when it reaches the satellite. (Warning: don't confuse radius and diameter.)

2. Suppose the beam of the lunar-ranging laser described in this chapter were expanded with a telescope to a 1-m waist. How wide would the beam be at the moon?

3. How wide is the laserbeam at the mirrors of the argon-ion laser described at the end of the section on Gaussian beams? Neglect any lensing introduced by the mirror, and calculate the beam radius 25 m from the laser.

4. Design a resonator for a HeNe laser that has a 0.25-mm waist on the output mirror and 30 cm between mirrors.

5. Suppose two TEM$_{00}$ HeNe lasers are sitting side by side and each has a flat output mirror. One laser has a 0.5-mm waist, and the other has a 1.0-mm waist. Obviously, the 0.5-mm beam will diverge more rapidly than the 1.0-mm beam, so at some distance from the laser both beams will have the same radius. What is that distance?

6. Which of the resonators sketched here are stable?

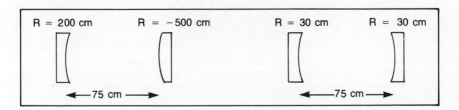

R = 200 cm   R = −500 cm   R = 30 cm   R = 30 cm

←——75 cm——→   ←——75 cm——→

7. What is the frequency spacing between adjacent longitudinal modes of the argon-ion laser described in question 3?

8. Suppose the mirrors of the laser in Fig. 9.14 were pulled apart. How would the frequency spectrum of Fig. 9.15 be changed? Sketch what the new spectrum would look like, showing any difference in the dotted curve and in the modes themselves. Repeat the sketch for the case when the mirrors are pushed together.

Ten

Reducing Laser Bandwidth

In this chapter we'll look at the ways to reduce a laser's bandwidth, that is, ways to make it even more monochromatic than it is naturally. A laser is the most nearly monochromatic optical source ever created, but for some applications—in precise spectroscopic studies or for separation of atomic isotopes, for example—a laser's natural bandwidth is just too large. Then the techniques discussed in this chapter are employed to reduce the bandwidth.

The concept of laser bandwidth was explained in Chapter 5. In this chapter we'll examine the different ways of measuring laser bandwidth and we'll discover the mechanisms that cause bandwidth. We'll have a look at the devices that are placed inside a laser resonator to reduce its bandwidth, and we'll see how a laser can be forced to oscillate in a single longitudinal mode.

Incidentally, there are several terms in laser technology for laser bandwidth. *Bandwidth, linewidth,* and *spectral width* all mean exactly the same thing: the degree of monochromaticity of the laser's output. And, as was explained in Chapter 5, the greater the *temporal coherence* of a laser, the smaller its bandwidth.

Measuring laser bandwidth

Because the bandwidth of a laser is such an important parameter, it's imperative to have some way of quantifying it. In fact, there are several ways of placing a numerical value on a laser's bandwidth: it can be measured in wavelength, or in frequency, or in wave numbers, or in coherence length.

A laser bandwidth measured in wavelength is shown in Fig. 10.1. In this case the neodymium laser's peak output is at 1.064 nm, but there's also some light at slightly shorter and slightly longer wavelengths. Notice that the bandwidth measurement is made *halfway* down from the peak of the laser line. This bandwidth is the "full-width, half-maximum" (FWHM) measurement, and it's the most common measurement of a laser's width.

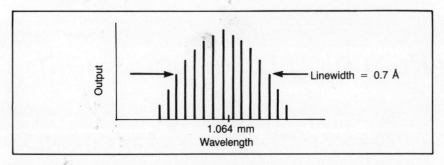

Fig. 10.1 *Laser output vs wavelength, showing longitudinal-mode structure*

But light can also be measured in frequency, so the bandwidth of a laser can be described in frequency, as shown in Fig. 10.2. Although this figure looks like Fig. 10.1, it shows that the laser peaks at 2.8×10^{14} Hertz (Hz), but there's some light at slightly higher and slightly lower frequencies. Notice that the measured bandwidth is again the FWHM value.

Wave numbers are yet another dimension for measuring laser bandwidth—one that's left over from the early days of spectroscopy and is still in general use. The frequency of optical transitions used to be measured in wave numbers. When a spectroscopist said a transition occurred at 20,000 wave numbers, he meant that 20,000 of the optical wavelengths would fit into 1 cm, or that the wavelength was 1/20,000 of a centimeter (500 nm). The value is written 20,000 cm^{-1}, meaning "20,000 wavelengths per centimeter," but it is pronounced "20,000 wave num-

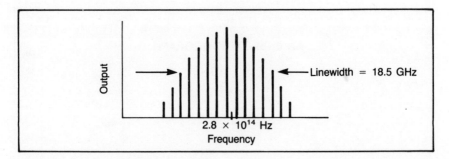

Fig. 10.2   *Laser output vs frequency*

bers" or sometimes "20,000 inverse centimeters." And since the frequency of light can be measured in wave numbers, the bandwidth can, too.

Fig. 10.3 is a nomograph that is useful not only in converting among bandwidth measurements but also among line-center measurements. The line-center conversions are made with the left side of the nomograph, and the bandwidth conversions are made with the right side. Down the middle of the nomograph is a column showing laser wavelength. The example shown is for a laser that has a wavelength of 532 nm and a bandwidth of 0.5 cm^{-1}. By following a horizontal line across the columns to the left, you can see that this laser has a line-center frequency of 5.6×10^{14} Hz or 18,800 cm^{-1}, or a photon energy of 2.3 electron-volts. (There are 6.2×10^{18} electron-volts in 1 joule.) On the right side of the nomograph, if you draw a straight, angled line connecting the laser wavelength with its bandwidth in the known dimension, you can read its bandwidth in the other dimensions. For the laser in the example, the bandwidth of 0.5 cm^{-1} corresponds to 15 gigahertz (GHz) or 0.14 angstroms (Å).

A fourth measurement of a laser's bandwidth is its coherence length. This is the distance over which the laser remains sufficiently coherent to produce interference fringes. It is inversely proportional to the laser bandwidth expressed in frequency or wavelength, and it is equal to the reciprocal of the bandwidth in wave numbers.

Laser broadening mechanisms

Why does a laser have a finite bandwidth?

Laser bandwidth derives from the "fuzziness" of the energy levels involved in the stimulated transition. Energy levels of a collection of

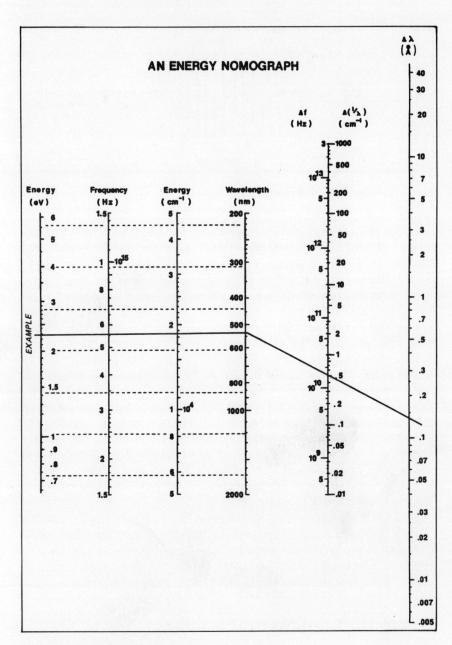

Fig. 10.3 *An energy nomograph*

atoms or molecules aren't razor-sharp, like those in Fig. 10.4a, but have a definite width to them, as illustrated in Fig. 10.4b. So the photons emitted when atoms (or molecules) undergo a transition won't all have exactly the same energy or the same wavelength.

There are several mechanisms that contribute to the width of energy levels. First, let's look at gas lasers, which have different broadening mechanisms than solid-state lasers. The atoms (or molecules) in a gas laser are free to bounce around inside the laser tube, while the atoms in a solid-state laser are pretty well tied down at a particular spot.

Doppler broadening is significant in almost all gas lasers. You experience the *acoustic* Doppler effect when a car blows its horn as it speeds past you. As the car approaches, the horn sounds high-pitched because the sound source and the sound wave itself are both coming toward you. This means that more waves per second enter your ear than if the car were standing still. When the car moves away from you, on the other hand, the source is moving away from you while the wave is moving in the opposite direction—toward you. So fewer waves per second enter your ear than if the car were standing still. What you hear is a low-pitched horn.

If you were standing in the middle of an intersection with many cars coming at you and moving away from you with different speeds, all blowing their horns, you'd hear a broad range of tones, even though all the horns produce the same tone when standing still. The sound from the cars moving away from you would be Doppler shifted down in pitch, while the cars moving toward you would be Doppler shifted up. The faster a car moved, the greater the Doppler shift would be. So you'd experience this broad range of tones, despite the fact that all the horns were actually vibrating at exactly the same frequency.

The *optical* Doppler effect increases the bandwidth of gas lasers. Because the individual atoms are moving about in random directions and at random speeds in the laser tube, their total emission covers a

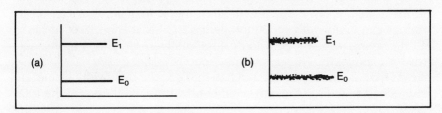

Fig. 10.4 *Energy levels aren't razor-sharp like (a); instead, they are slightly fuzzy like (b)*

range of frequencies just as the acoustic emission from the cars covered a range of frequencies. The faster the atoms move on the average—that is, the hotter the gas is—the broader the bandwidth. Moreover, there is a relativistic effect in the optical Doppler shift. A relativistic time dilation in the moving reference frame also contributes to the frequency shift.

In a Doppler-broadened laser, the bandwidth of an individual atom (or molecule) is smaller than the laser bandwidth. A single photon might be able to stimulate one atom to emit because that atom happened to be Doppler shifted to the photon's frequency, but it might not be able to stimulate another atom because it had a different Doppler shift than the first. This type of broadening, where the different atoms contribute to the gain at different frequencies within the laser bandwidth, is called *inhomogeneous broadening*. It's illustrated schematically in Fig. 10.5

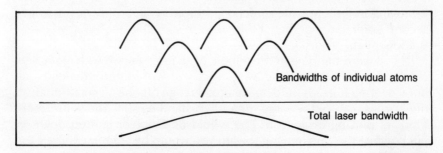

Bandwidths of individual atoms

Total laser bandwidth

Fig. 10.5 *In an inhomogeneously broadened laser, individual atoms emit at different frequencies*

Homogeneous broadening is illustrated in Fig. 10.6. In a homogeneously broadened laser, each individual atom has a bandwidth equal to the total laser bandwidth. If a particular photon can interact with one of the atoms, it can interact with all of them. In general, it's easier to reduce the bandwidth of a homogeneously broadened laser because all the atoms can still contribute to stimulated emissions at the narrower bandwidth. In an inhomogeneously broadened laser, those atoms that contribute to gain outside the reduced bandwidth cannot be stimulated to emit in the narrowed bandwidth, and therefore the total laser power is reduced.

An example of homogeneous broadening in a gas laser is *pressure broadening* (or *collision broadening*, as it's sometimes called). One result of the uncertainty principle of physics is that the natural band-

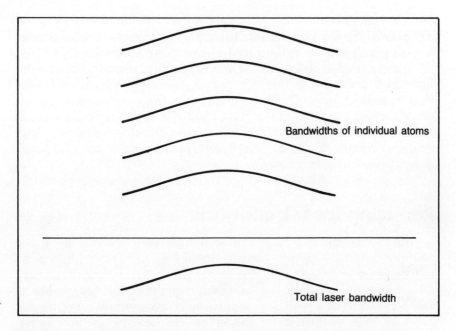

Fig. 10.6   *In a homogeneously broadened laser, all atoms are the same*

width of an atom is inversely proportional to the time between collisions. That is, the longer the atom can travel in a straight line without
bumping into something (like another atom or the side of the laser
tube), the narrower its natural bandwidth. It makes sense that the
fewer other atoms there are in the tube—that is, the lower the gas
pressure in the tube—the longer will be the average time between
collisions. So as the pressure in a laser tube increases, the bandwidth of
the laser also increases. This broadening mechanism is homogeneous
because it increases the total lasing bandwidth by increasing the bandwidths of the individual atoms.

Doppler broadening and pressure broadening are the most important broadening mechanisms in a gas laser. If the tube contains low-
pressure gas, Doppler broadening is predominant; at high gas pres-
sures, pressure broadening becomes more important.

In a solid-state laser the individual lasing atoms are tied down to the
host crystal's lattice points so they can't move around, bumping into
things and being Doppler broadened by their velocity. But there are
other broadening mechanisms in solid-state lasers, the most important
of which is thermal broadening. Although the atoms are attached to the
crystal lattice, the lattice itself is subject to vibration from thermal

energy. This vibration modulates the natural emission frequency of the atoms and thereby broadens it. Thermal broadening is homogeneous because each atom is subject to the same thermal vibration.

When a solid-state laser is operated at very low temperature and the thermal broadening is therefore small, residual broadening results from imperfections of the host crystal. These imperfections are different at various locations in the crystal and give rise to differing electric fields at the active atoms. These fields cause different frequency shifts in the different atoms, so crystal-field broadening is an inhomogeneous broadening mechanism.

Reducing laser bandwidth

The bandwidth of a laser can be narrowed by chilling the active medium to reduce thermal broadening (if it's a solid-state laser) or to reduce Doppler broadening (if it's a gas laser). But chilling isn't a very effective way of reducing the bandwidth, and it's often inconvenient. The bandwidth of a gas laser can usually be reduced by reducing the pressure, but unfortunately this often reduces the output power because there are fewer atoms left to lase. There is another way of reducing a laser's bandwidth, a way that does not cause a disastrous reduction of the laser's output power. Remember that the two conditions necessary for lasing are (1) the existence of a population inversion and (2) a round-trip gain greater than unity. The techniques mentioned above—chilling and pressure reduction—reduce laser bandwidth by reducing the bandwidth of the population inversion. The more effective techniques reduce laser bandwidth by reducing the bandwidth of the laser's round-trip gain. That is, the feedback of the resonator is modified to control the lasing bandwidth.

Suppose a laser has a population inversion that is 4 GHz wide and has mirrors that have reflectivity wider than the population inversion. This situation is diagramed in Fig. 10.7a. Because all the light within the bandwidth of the population inversion sees round-trip gain greater than unity, the laser lases over the entire population inversion and the output bandwidth is 4 GHz.

But if the mirrors were replaced by special mirrors with a bandwidth of only 1 GHz, then only part of the light within the bandwidth of the population inversion would see the round-trip gain necessary for lasing. Lasing could occur only in this narrow band, and the bandwidth of the output would be reduced, as diagramed in Fig. 10.7b.

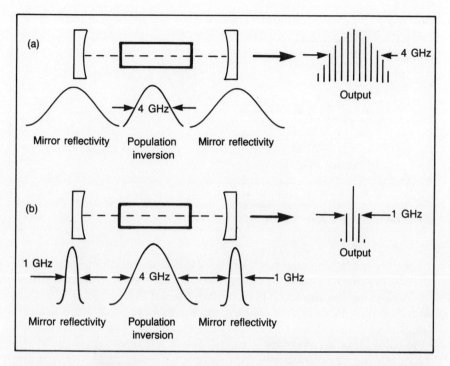

Fig. 10.7 (a) If the resonator provides wide-band feedback, then the bandwidth of the output will be as large as the bandwidth of the population inversion; (b) but if the bandwidth of the feedback is reduced, the laser bandwidth will be, too

That's the fundamental approach to reducing the bandwidth of a laser: You must reduce the bandwidth of the resonator feedback. The approach of Fig. 10.7b isn't very practical because it is difficult, if not impossible, to make laser mirrors with a 1-GHz bandwidth. So other devices are used to reduce the resonator feedback, but the principle is always exactly what's shown in Fig. 10.7b.

An intracavity prism is one common device that reduces the bandwidth of feedback in a resonator. The idea is shown in Fig. 10.8. Although the population inversion is 4 GHz wide, only light at the center of this bandwidth is bent directly toward the mirror by the prism. Light at the edges of the population-inversion width—λ_1 and λ_3 in Fig. 10.8—emerges from the prism at a different angle and can't be reflected back by the mirror. So this resonator produces round-trip

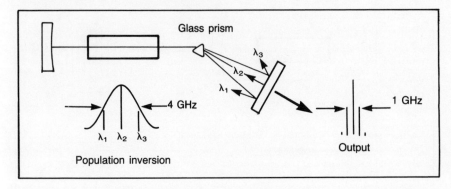

Fig. 10.8 *An intracavity prism reduces the bandwidth of resonator feedback and, therefore, the laser bandwidth*

gain only for the narrow band of light at the center of the population inversion, and laser output is restricted to this reduced bandwidth.[1]

An alternative approach is to replace one of the mirrors with a grating. A grating is an interferometric device that reflects different wavelengths at different angles. When it's aligned correctly at one end of a resonator, it will reflect back to the active medium only light at the center of the population inversion, as shown in Fig. 10.9. So the lasing bandwidth is again reduced to the bandwidth of the resonator's round-trip gain.

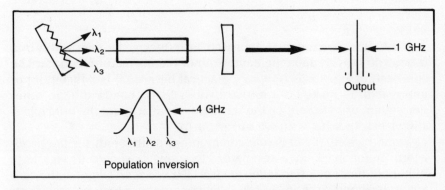

Fig. 10.9 *An intracavity grating will also restrict a laser's bandwidth*

[1]*Several prisms in a series are necessary to reduce the bandwidths of some lasers.*

What would happen if you put the bandwidth-reducing device outside the laser resonator? For example, Fig. 10.10 shows a prism in the output beam of a laser and an aperture that passes only a narrow bandwidth. At first glance, you might think this arrangement preferable because the straight-line resonator would be more easily aligned than a bent resonator. What's the catch?

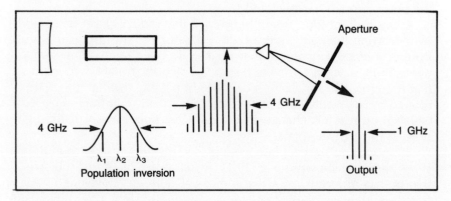

Fig. 10.10   *One technique to obtain narrow-bandwidth light from a laser*

The catch is that you're losing most of the laser light in Fig. 10.10. All the light that hits the edge of the aperture is lost, and the only useful output is the small fraction that passes through the aperture. When the bandwidth-limiting device is inside the laser, however, nearly as much laser power can be produced in a narrow bandwidth as in a wide bandwidth.

You can do better in a homogeneously broadened laser than in an inhomogeneously broadened laser because every atom in the population inversion can still contribute to the laser output. (In Fig. 10.6, every atom can be stimulated by light at the center frequency.) In an inhomogeneously broadened laser, some of the atoms are unable to contribute to the reduced-bandwidth output and the laser power is reduced. (In Fig. 10.5, some atoms cannot be stimulated by light at the center frequency.) But even in an inhomogeneously broadened laser, it's better to do the bandwidth reduction inside the resonator because some of the atoms outside the lasing bandwidth may eventually contribute to the laser gain. For example, they could collisionally transfer their energy to atoms that can emit within the lasing bandwidth.

Based on what's been said so far, it might seem that a homogeneously broadened laser could be restricted to a narrow bandwidth with no loss of output power—because as many atoms can contribute to the narrow-band output as to the wide-band output. In reality, it doesn't work out that way for several reasons. For one thing, the insertion of an extra element, such as a prism, into a laser resonator always causes some reduction in output because there's no such thing as a perfect (lossless) optical element. An effect called *spatial hole burning,* which will be discussed in the next section, also plays a role in lowering the power from a narrow-band laser.

A birefringent filter is another device that reduces laser bandwidth by narrowing the bandwidth of the resonator's round-trip gain. Recall from problem 5, Chapter 3, that a half-wave plate retards one component of polarization 180° with respect to its orthogonal component so that the effective polarization of the light passing through the plate is rotated by 90°. A full-wave plate, then, retards one component by 360°, which produces no change in the effective polarization of the light passing through the plate.

Suppose a full-wave plate were placed inside a resonator at Brewster's angle, as shown in Fig. 10.11. Because the plate is dispersive—that is, has different refractive indices at different wavelengths—it's exactly a full-wave plate at only one wavelength. Light at slightly offset wavelengths will experience a slightly different retardation, maybe 359° or 361°. So the light at the offset wavelengths will be slightly elliptically polarized after passing through the plate. Now look at Fig. 10.11 again. The plate at Brewster's angle will reflect part of the elliptically polarized light out of the resonator, but light at the central wavelength will be perfectly plane polarized and won't be reflected out of the cavity. So the round-trip gain for light at the offset wavelengths will be lower than that for the central wavelength. If it's enough lower, the bandwidth of the laser will be reduced.

In practice, a birefringent filter is usually made up of a full-wave plate, a two-full-wave plate, and sometimes a three-full-wave plate, all

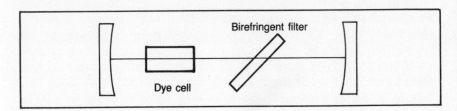

Fig. 10.11 *A birefringent filter inside a dye-laser resonator*

assembled together in a single unit. The additional waveplates decrease the bandwidth of the device.

Single-mode lasers

Each transverse mode and each longitudinal mode of a laser oscillates at a different frequency, and in an unrestricted laser numerous modes of both types oscillate simultaneously. In Chapter 9 we saw that an intracavity aperture can force a laser to oscillate in a single transverse mode. The ultimate narrow-bandwidth laser oscillates in only a single transverse and longitudinal mode. Normally, the techniques introduced in the previous section aren't restrictive enough to force a laser to a single mode.

If an aperture is placed in the resonator of an otherwise-unrestricted laser, the laser will oscillate in a comb of frequencies corresponding to the different longitudinal modes of a single transverse mode, as shown in Fig. 10.12. In this figure, the laser output at any frequency is determined by the product of the laser gain, the mirror reflectivity, and the resonator mode structure. If you want to know the output from the laser at a particular frequency, you must multiply together the gain, reflectivity, and mode structure at that frequency. For some frequencies the product is zero because the mode structure is zero, so there is no output at those frequencies.

If the bandwidth of the laser is reduced, perhaps with a prism, fewer modes oscillate. This situation is diagramed in Fig. 10.13, where the output is again shown as determined from the product of population inversion and feedback components.

To restrict a laser to a single mode, it's usually necessary to place an *etalon* inside the resonator. An etalon is nothing more than two surfaces that act like a Fabry-Perot interferometer. Remember that the transmission peaks of a Fabry-Perot are separated by $c/2L$, where L is the distance between the reflecting surfaces. If this distance is small, then there's a relatively large frequency spacing between adjacent transmission peaks. These peaks (which are the longitudinal modes of the etalon) act in concert with the longitudinal modes of the resonator to extinguish all but one of the laser's longitudinal modes, as shown in Fig. 10.14.

In practice, an etalon often is a piece of optical-quality glass fabricated with great care to ensure that the surfaces are parallel. The surfaces may be coated to enhance reflectivity, or they may be uncoated.

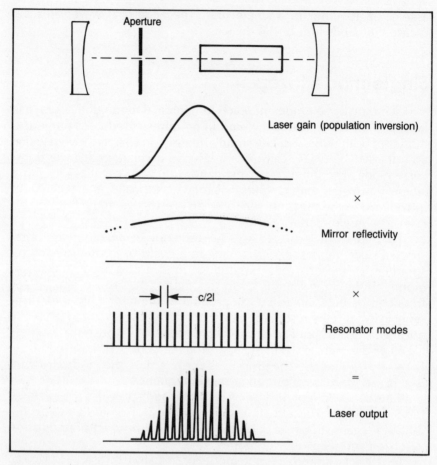

Fig. 10.12 *A laser restricted to a single transverse mode can oscillate in many longitudinal modes*

(Reflective coatings increase the etalon's *finesse*, the ratio of transmission separation to width. Coatings would be required to achieve the relatively high etalon finesse shown in Fig. 10.14.) Many single-mode lasers use more than one etalon to make sure the laser is restricted to one mode.

A single-mode laser is often called a single-frequency laser. But a single-wavelength laser is something else. The term usually refers to a laser, like an ion laser, that can lase on more than one transition but has been artificially restricted to one transition.

Suppose that you had a perfect, lossless etalon—one whose flawless surfaces scattered absolutely no light and whose magical bulk

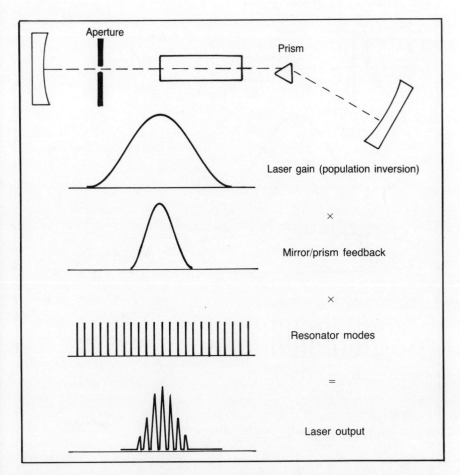

Fig. 10.13 *Fewer modes oscillate when the bandwidth of resonator feedback is restricted*

material absorbed absolutely none. If you placed this device inside a homogeneously broadened laser, you might expect to get as much output from the resultant single mode as you'd been getting from all the longitudinal modes put together. After all, every atom that contributed to the output before could still contribute to the single-mode output, right?

Wrong. To see why, look at the spatial distribution of electrical fields in a single longitudinal mode, as diagramed in Fig. 10.15. At the *nodes* of the standing wave, there is no electrical field, so the atoms located right at the nodes can't be stimulated to emit their energy. In fact, the single mode will "burn holes" in the population inversion at

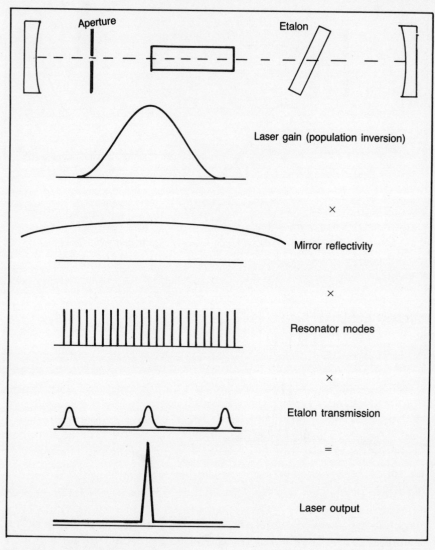

Fig. 10.14 *Single-mode oscillation*

those locations where the electric field is greatest. And the atoms at the nodes of the single mode can't contribute to its output, even though they can emit at the correct frequency.

In practice, a single etalon usually will not force a homogeneous laser to oscillate in a single mode. The gain from atoms located at nodes

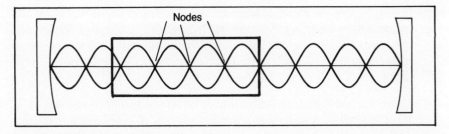

Fig. 10.15 *Atoms at nodes of standing waves cannot contribute to laser gain*

of the preferred mode becomes so great that one or more additional modes will oscillate despite the etalon. You can force single-mode oscillation by adding a second (and maybe a third) etalon to the resonator, but only at the cost of reduced output power. The output power is reduced because the atoms located at the nodes of the single mode cannot be tapped for their energy.

Questions

1. Find the optical frequency of an argon laser whose wavelength is 514.5 nm from the nomograph (Fig. 10.3) and from direct calculation. Suppose the laser has a 0.4 cm^{-1} bandwidth. Calculate the bandwidth in frequency and wavelength, and confirm your calculation from the nomograph.

    An argon laser can also produce a blue line at 488 nm. If it does and its bandwidth is still 0.4 cm^{-1}, what is its bandwidth in frequency and wavelength?

2. Spectral intensity is a measure of laser power within a given bandwidth. For example, a 1-watt laser with a 1-angstrom bandwidth has a spectral intensity of 1 W/Å. What's the spectral intensity of a 500-mW Nd:YAG laser whose bandwidth is 0.3 Å?

    Suppose a filter *outside* the laser reduces the bandwidth to 0.1 Å but reduces the power to 170 mW. What's the spectral intensity now?

    Suppose a prism *inside* the laser reduces the bandwidth to 0.1 Å but reduces the power to 450 mW. What's the spectral intensity in this case?

    Which is the better way to obtain narrow-band output from the laser?

3. It is possible to force a laser to oscillate in a single longitudinal mode just by pushing the mirrors close enough together. Why does this work? How close must you push the mirrors of a laser whose unrestricted bandwidth is 1.0 GHz to force it into single-mode oscillation?

4. Consider a dye laser whose output has been tuned to 600 nm. The laser is analyzed with a scanning spherical-mirror interferometer, and its spectral width is determined to be 0.5 Å. Express this spectral width in (a) frequency and (b) wave numbers.

   If the laser cavity is one meter long, about how many longitudinal modes can oscillate?

Eleven

Q-switching

Q-switching is the first of several techniques that we'll examine for producing pulsed output from a laser. Pulsed lasers are useful in many applications where continuouswave (cw) lasers won't work because the energy from a pulsed laser is compressed into concentrated little packages. This concentrated energy in a laser pulse is more powerful than the natural-strength energy that comes from a continuouswave laser.

We'll begin this chapter by looking at the way the output of a pulsed laser is measured. It's more complicated than simply measuring the average power output of a cw laser. Then the concept of Q-switching—storing up energy inside the laser and suddenly letting it all out in a giant pulse—will be explained in detail. The chapter will conclude with a discussion of the four different types of Q-switches that can make a laser store its energy for emission in pulses.

Measuring the output of pulsed lasers

Measuring the output of a cw laser is pretty simple because the energy flows smoothly and constantly from the laser, as shown in Fig. 11.1. But with a pulsed laser, you want to know the answers to other questions. Are there a lot of little pulses or a few big ones? And how tightly is the energy compressed in the pulses?

When you measure the output of a cw laser, you measure the amount of energy that comes out during a given period of time. The energy is measured in a dimension called joules, and time is measured in seconds. The *rate* at which energy comes from the laser—that is, the number of joules per second (J/s)—is the power of the laser, measured in watts (W).

Thus, to a physicist the words "power" and "energy" have different meanings. Energy is measured in joules and is defined as the ability to do work (like moving or heating something). Power, on the other hand, is the rate of expending energy and is measured in joules per second (watts). For example, a 100-W lightbulb uses up 100 J of electrical energy every second it's on. If you leave it on for 5 minutes, you've used up 30,000 J.

When your electricity bill comes, is it based on the *power* you've used that month or on the *energy*? Clearly, it doesn't make sense to talk about how much power you've used during a month because power is the rate of using something. So your electricity bill, for so many kilowatt-hours, is a bill for energy. You use a kilowatt-hour of electrical

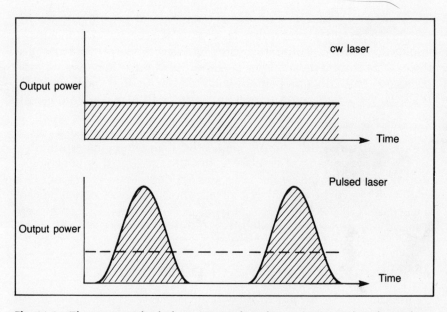

Fig. 11.1 *The energy (shaded area) is produced in concentrated packages by a pulsed laser. Because the energy is concentrated, its peak power is greater*

energy when you expend energy at the rate of 1 kW for 1 hour. Leaving ten 100-W bulbs on for an hour would do it, or you could use a 4-kW clothes dryer for 15 minutes. You've used a kilowatt of power (1,000 J/s) for one hour, or you've used a total of 3,600,000 J. A kilowatt-hour of electricity costs a little less than a dime, depending on where you live, so you can see that a joule isn't a whole lot of energy.

There are two power measurements for a pulsed laser: the *peak power* and the *average power*. The average power is simply a measurement of the average rate at which energy flows from the laser during an entire cycle. For example, if a laser produces a single half-joule pulse per second, its average power is 0.5 W. The peak power, on the other hand, is a measurement of the rate at which energy comes out during the pulse. If the same laser produces its half-joule output in a microsecond-long pulse, then the peak power is 500,000 W (0.5 joules/10^{-6} seconds = 500,000 J/s).

The *pulse repetition frequency* (prf) is a measurement of the number of pulses the laser emits per second. The *period* of a pulsed laser is the amount of time from the beginning of one pulse until the beginning of the next. It is the reciprocal of the prf. The *duty cycle* of a laser is the fractional amount of time that the laser is producing output, the pulse duration divided by the period.

As an example, let's consider a flash-pumped, Q-switched Nd:YAG laser that produces 100-millijoule (mJ), 20-ns pulses at a prf of 10 Hertz (Hz). The average power from this laser is 1 W, the peak power is 5 megawatts (MW), the period is 0.1 second, and the duty cycle is 2 × 10^{-7}. Be sure you understand this calculation before continuing.

Q-switching

Q-switching is a simple concept. Energy is stored in the population inversion until it reaches a certain level; then it's released very quickly in a giant pulse. This is analogous to storing water in a flower pot with a hole in the bottom and then releasing the water all at once, as shown in Fig. 11.2.

Now the question is: How can energy be plugged up in the population inversion of a laser? That is, how can you prevent the energy from draining out of the population inversion as fast as it goes in?

To prevent the laser from lasing, you must defeat one of the two requirements for lasing: you must eliminate either the population inversion or the feedback. Obviously, if energy is stored in the population inversion, it doesn't make sense to talk about eliminating that.

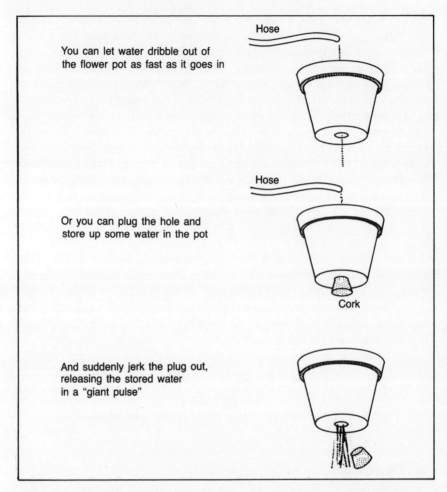

You can let water dribble out of the flower pot as fast as it goes in

Hose

Or you can plug the hole and store up some water in the pot

Hose

Cork

And suddenly jerk the plug out, releasing the stored water in a "giant pulse"

Fig. 11.2 *Storing up water in a flower pot is analogous to Q-switching a laser*

But you can eliminate feedback, thereby preventing lasing and thus storing up all the extra energy in the population inversion, by blocking one of the laser mirrors.

That's exactly the way a laser is Q-switched. As shown in Fig. 11.3, if the normal mirror feedback is present, energy drains out of the population inversion as fast as it's put in. But if feedback is eliminated, energy builds up in the population inversion until feedback is restored, and then all the energy comes out in a single, giant pulse.

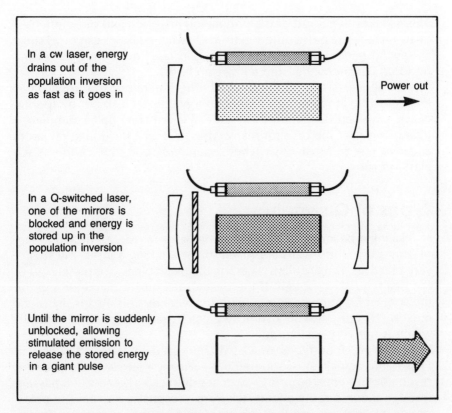

Fig. 11.3 *Energy is stored in the population inversion of a Q-switched laser. This diagram shows a lamp-pumped solid-state laser*

Why is it called "Q-switching"? The Q stands for the "quality" of the resonator. A high-Q resonator is a high-quality resonator, or one that has low loss. Obviously, a resonator with a blocked mirror isn't very high Q. But when the mirror is suddenly unblocked, the Q is *switched* from low to high. Thus, a Q-switched laser is one whose resonator can be switched from low quality to high quality and back again.

It's interesting to note that the Q-switch must switch the resonator from low Q to high Q to start the pulse, but stimulated emission will use up all the population inversion and terminate the pulse. That is, the laser lases until the population inversion is all gone, not until the Q-switch turns it off. The Q-switch has very little to do with how long the pulse lasts. In fact, pulse duration depends mainly on how long the energy takes to get out of the population inversion and out of the resonator itself. The pulse duration from Q-switched lasers can be as short as several nanoseconds from a high-gain laser or as long as several hundred nanoseconds from a low-gain laser.

Not all lasers can be Q-switched. The lifetime of the upper laser level must be long enough so that energy doesn't leak out by spontaneous emission before the resonator Q is switched up to stimulate a pulse. This isn't the case for many types of lasers, including ion lasers and dye lasers. Most Q-switched lasers are solid-state, either YAG, glass, or ruby.

Types of Q-switches

Placing a beam block in front of the laser mirror, as shown in Fig. 11.3, is a straightforward approach to Q-switching a laser, but it isn't very practical. The problem is getting the beam block completely out of the beam quickly enough. If the beam is 0.5 mm in diameter and the block must be pulled out in a few nanoseconds, this means the block must be jerked out with a velocity greater than the speed of sound—which is not very easy to do.

There are four types of Q-switches used in lasers. *Mechanical* Q-switches actually move something—usually a mirror—to switch the resonator Q. *Acousto-optic* Q-switches diffract part of the light passing through them to reduce feedback from a resonator mirror. The polarization of light passing through an *electro-optic* Q-switch can be rotated so that a polarizer prevents light from returning from a mirror. And a *dye* Q-switch absorbs light traveling toward the mirror until the intensity of the light becomes so great that it bleaches the dye, allowing subsequent light to pass through the Q-switch and reach the mirror.

Mechanical Q-switches

A mechanical Q-switch is shown in Fig. 11.4. Here, the six-sided mirror spins rapidly and lines each side up with the laser for a very short period. A laser pulse comes out through the other mirror each time one of the six sides of the spinning mirror is aligned. These rotating-mirror Q-switches were fairly common in the early days of lasers (the early 1960s), but other types of Q-switches have replaced them for most applications now.

Another type of mechanical Q-switch is the frustrated total-internal-reflection (FTIR) Q-switch. This device provides feedback to the laser

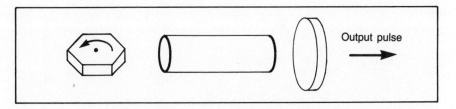

Fig. 11.4 *The quality of this resonator is switched from low to high when a surface of the spinning mirror is aligned with the other mirror*

by total internal reflection from the inner surface of a prism. To reduce resonator Q, a second prism is pushed quickly into optical contact with the reflecting surface of the first prism, frustrating the total internal reflection. These FTIR Q-switches aren't very common either, but they do find applications in special situations.

Although mechanical Q-switches are conceptually simple, they have several drawbacks that preclude their general use. Because they require rapidly moving mechanical parts, the long-term reliability of mechanical Q-switches is poor. Also, it's difficult to synchronize external events with the pulse from a mechanically Q-switched laser. This task can be accomplished, for example, by aligning a small HeNe laser and detector with a spinning mirror Q-switch so that the detector produces a signal just before the laser pulse is emitted. But this is awkward and inconvenient and is subject to misalignment.

Acousto-optic Q-switches

An acousto-optic (A-O) Q-switch is a block of transparent material, usually quartz, with an acoustic transducer bonded to one side. This transducer is similar to a loudspeaker because it creates a sound wave in the transparent material, just as a stereo speaker produces a sound wave in your living room (except most A-O Q-switches operate at ultrasonic frequencies). Now this sound wave is a periodic disturbance of the material, and any light that happens to be traveling through the material sees this periodic disturbance as a series of slits, just like those in Young's double-slit experiment (see Chapter 4 if you don't remember exactly how Young's experiment worked). Thus, the light is diffracted out of the main beam by interference, as shown in Fig. 11.5.

The idea, then, is to place an acousto-optic Q-switch inside a laser between the gain medium and the back mirror, as shown in Fig. 11.6. If no acoustic signal is applied to the transducer, then the Q-switch trans-

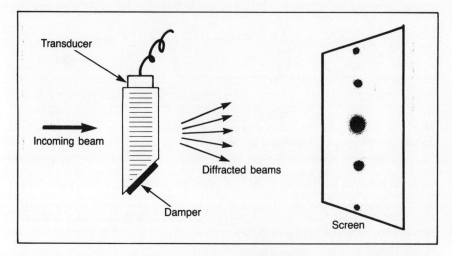

Fig. 11.5 *The incoming beam of light is diffracted from the periodic disturbance of the sound wave in an acousto-optic Q-switch*

mits all the light without disturbing it and the resonator has a high Q. But when an acoustic signal is applied to the transducer, light is diffracted out of the intracavity beam and the resonator Q is reduced.

The side of the acousto-optic Q-switch opposite the transducer is usually configured to minimize reflection of the acoustic wave, as shown in Fig. 11.5. A damping material absorbs most of the sound wave's energy, and what isn't absorbed is reflected back off-axis by the oblique surface. If the reflected wave weren't minimized, it could interfere with the original wave and reduce the diffraction efficiency of the Q-switch. (Acousto-optic modelockers, on the other hand, operate at much higher frequencies and depend on acoustic waves traveling in both directions; see Chapter 12.)

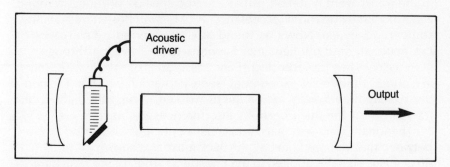

Fig. 11.6 *An acousto-optic Q-switch placed inside a laser*

The speed of an acousto-optic Q-switch—that is, how quickly it can switch the resonator from low Q to high Q—depends on the sound velocity within the block of transparent material and on the diameter of the laser beam. After all, the switching time is simply the time it takes the sound wave to get out of the way of the beam. So the smaller the intracavity laser beam is, the faster the speed of the Q-switch.

Acousto-optic Q-switches are frequently used in lasers because they are less expensive than electro-optic Q-switches and their speed is good enough for many applications. Acousto-optic Q-switches are easy to synchronize with other events because the pulse is emitted with a constant delay after an acoustic signal is applied to the transducer. The main drawback of acousto-optic Q-switches is their low *hold-off*, that is, their limited ability to keep a high-gain laser from lasing. Only part of the light is diffracted when it passes through the Q-switch, and the remainder is fed back to the laser. If laser gain is great enough, this small feedback can be enough to make the round-trip gain greater than the round-trip loss, and the laser will lase. So acousto-optic Q-switches can be used only with low-gain lasers.

Electro-optic Q-switches

An electro-optic (E-O) Q-switch has two elements, as shown in Fig. 11.7. The Pockels cell is an electrically controlled wave plate; light passes through the Pockels cell unchanged when no voltage is applied to the cell. But when the Pockels cell is biased, it causes the polarization of light passing through to be rotated 90°. The second element of the Q-switch is a polarizer that passes the light in its nonrotated polarization but rejects the rotated polarization.

An electro-optic Q-switch is placed inside a laser resonator between the back mirror and the gain medium, as shown in Fig. 11.8. As long as a voltage is applied to the Pockels cell, resonator Q remains low and energy builds up in the population inversion. When the voltage is switched off, resonator Q is restored and all the stored energy is emitted in a Q-switched pulse. Incidentally, it is important that the Q-switch be placed between the *back* mirror and the gain medium because the back mirror has greater reflectivity than the front mirror. Thus, the Q-switch is positioned to block the more effective mirror when it's placed between the back mirror and the gain medium.

Now we have to take a closer look at what happens inside the Pockels cell. How does it rotate the polarization of light passing through it?

The Pockels cell is made of a material that displays the electro-optic effect—its refractive index can be changed by an electric field. More-

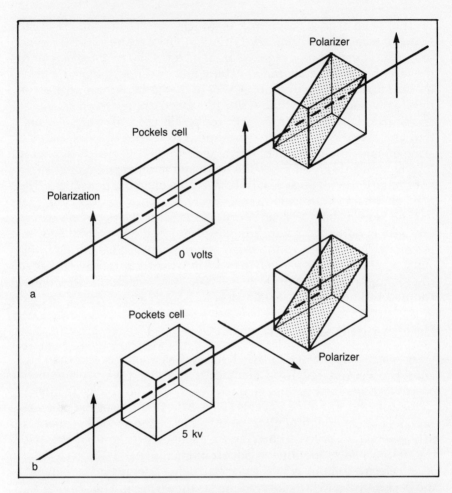

Fig. 11.7   *When voltage is applied to a Pockels cell, polarization of light passing through it is rotated and the light can be deflected from a subsequent polarizer*

over, the material must also be *birefringent,* so its refractive index depends on the polarization of light passing through it.

To understand how a Pockels cell works, recall that linearly polarized light can be thought of as having two orthogonally polarized, in-phase components (review Chapter 3 if you're hazy on this). When these two components pass through an unbiased Pockels cell, as shown in Fig. 11.9a, both components see the same refractive index and therefore move through the cell at the same velocity. The light that emerges from an unbiased Pockels cell has the same polarization as the incoming light.

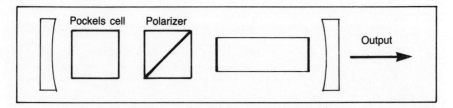

Fig. 11.8 *An electro-optic Q-switch is placed in a resonator between the back mirror (maximum reflectivity) and the gain medium*

But when a voltage is applied to the Pockels cell, the refractive indices are changed (the electro-optic effect) and the two components no longer see identical refractive indices. Because they see different refractive indices, the two components move at different velocities through the Pockels cell. So one component gets ahead of the other, and when they emerge they don't have the phase relationship they started out with. In fact, if the velocity difference and Pockels-cell length are just right, the phase between the two will have changed exactly 180° (one-half wave), as shown in Fig. 11.9b. If you'll add these two components together, you'll find that the light emerging from the Pockels cell is polarized orthogonally to the incoming light. That is, by retarding one polarization component 180° with respect to the other, the Pockels cell rotated the actual polarization of the light by 90°.

In practice, an electro-optic Q-switch is usually placed inside a laser, as shown in Fig. 11.8, so that light from the gain medium makes *two* passes through the Pockels cell before returning to the polarizer. This way the Pockels cell need only retard one component by a quarter-wave each time, and the voltage that must be applied to the Pockels cell is accordingly reduced.

An electro-optic Q-switch is probably the most effective of the four types of Q-switches. There are no moving parts—not even a sound wave—so it's very fast and reliable. It's also easy to synchronize because the laser pulse comes out immediately after the voltage on the Pockels cell is switched. But an electro-optic Q-switch is very expensive. The electro-optic materials from which it is made (usually potassium dihydrogen phosphate or one of its isomorphs) are expensive and the power supplies to drive them are quite costly. Quarter-wave voltages for most Pockels cells are just over 1 kV, and switching times are generally a few nanoseconds. Switching voltages that high in times that short requires expensive electronics.

A Kerr cell is similar to a Pockels cell, except a liquid medium instead of a crystal provides the phase retardation. Nitrobenzene is the liquid most commonly used, and the voltage that must be applied is

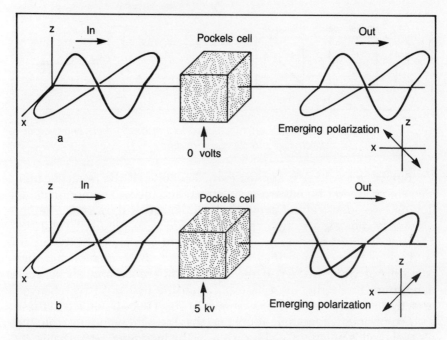

Fig. 11.9 *(a) When no voltage is applied to a Pockels cell, the material displays no birefringence and light passing through is not rotated. (b) When the Pockels cell is biased, one component moves through faster than the other so the polarization of the emerging light is rotated*

much greater than what's required for a Pockels cell. Optical damage of a Kerr cell tends to be self-healing. But because they're messy and require very high voltages, these devices are seldom used.

Dye Q-switches

Dye Q-switches, also called saturable-absorber Q-switches or passive Q-switches, utilize a dye whose transmission depends on incident light intensity, as shown in Fig. 11.10. A cell containing this dye is placed inside the laser and blocks a mirror, as shown in Fig. 11.11. But when light emitted from the gain medium becomes intense enough (from both spontaneous emission and stimulated emission), the dye bleaches and light passes through it with little loss.

Dye Q-switches are inexpensive because of their simplicity, but they have several drawbacks, including pulse jitter, dye degradation, and synchronization difficulties. Nonetheless, they find frequent appli-

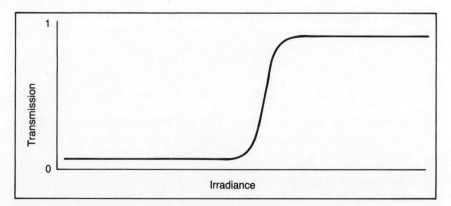

Fig. 11.10 *The transmission of a saturable absorber increases steeply beyond a certain irradiance*

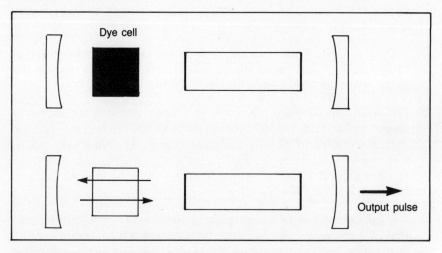

Fig. 11.11 *Photons emitted by spontaneous and stimulated emission bleach the dye Q-switch so light can pass through it to the mirror*

cation because they're so simple to use. Dye Q-switches are generally not used with low-gain (cw-pumped) lasers.

Questions

1. Think about a cw laser that produces a 1-W output. If this laser is Q-switched with no energy loss and produces 10 pulses per second of 100-ns duration, what is the peak power in each pulse?

2. The Q (quality) of a resonator is defined as:

$$Q = \frac{\text{energy stored in resonator}}{\text{energy lost/cycle}}$$

What's the Q of a Nd:YAG resonator that has 5% round-trip loss?

3. Calculate the switching speed of an acousto-optic modulator if the sound velocity within the modulator is 6×10^3 m/s and the intracavity beam of the laser is 0.5 mm in diameter.

4. Consider a chromium-ruby laser with 2×10^{19} chromium ions that contribute to the laser action. If each ion emits only one photon per pulse, what is the fractional population inversion required to produce a 1-J pulse? (The fractional population inversion is defined as $F = (N_1 - N_0)/(N_1 + N_0)$, where N_1 and N_0 are the populations of the upper and lower laser levels, respectively. Remember that Cr:ruby is a three-level system.)

5. Suppose a Nd:YAG laser is Q-switched with a rotating-mirror Q-switch. If the six-sided mirror rotates at 3,200 rpm, calculate the pulse repetition frequency (in Hertz) of the laser.

6. When an electro-optic Q-switch is placed inside a resonator as shown in Fig. 11.8, the light passes through the Pockels cell once in each direction and its polarization is rotated 90°. What's the polarization of the light after a single pass through the Pockels cell? (That is, what's the polarization of the light between the Pockels cell and the back mirror?)

7. The pulse duration of a high-gain, Q-switched laser is shorter than that of a low-gain laser because high gain tends to empty out the population inversion more quickly. Pulse duration also depends on two other parameters: resonator length and mirror transmission. What effect would each of these parameters have on the pulse duration and why?

Twelve

Cavity Dumping and Modelocking

Pulsed output can be obtained from many lasers by Q-switching, but this technique won't work with lasers whose upper-state lifetime is too short to store appreciable energy. Another approach—cavity dumping—must be used to obtain pulses from these lasers. And even in lasers that can be Q-switched, there's a lower limit on pulse duration and an upper limit on pulse repetition frequency imposed by the natural time constants of the population inversion. Cavity dumping also can be used to obtain very short and/or very high frequency pulses from these lasers.

But the highest pulse frequencies, and the shortest pulses, are obtained by modelocking a laser. Any type of laser can be modelocked—even one that's already been Q-switched or cavity dumped. In fact, some lasers are simultaneously Q-switched, cavity dumped, and modelocked.

Cavity dumping

Cavity dumping is a very descriptive name for the process of obtaining a pulsed output from a laser, if you understand that the word *cavity* is used to mean "resonator." One type of cavity-dumped laser is shown in Fig. 12.1. Notice that both mirrors are maximum reflectors; neither transmits any light. So how does anything get out of the laser? Let's examine what happens when this laser produces a cavity-dumped pulse.

When the lamp flashes, the electro-optic Q-switch is in its "transmit" mode; that is, light passes through the Pockels cell without any polarization rotation, so the polarizer doesn't reject the light. As soon as the round-trip gain becomes equal to the round-trip loss, the laser begins lasing and light starts bouncing back and forth between the mirrors. Because neither mirror transmits any light, this circulating power builds up to a high level. When the maximum intracavity circulating power has been obtained, a voltage is applied to the Pockels cell. Now the cell rotates the polarization of light passing through it, and the polarizer ejects all the light in a cavity-dumped pulse, as shown in Fig. 12.2.

So the output coupler for a cavity-dumped laser is the cavity dumper itself and not a mirror, as has been the case for all the other lasers we've examined so far. Later we'll see additional cases where the output coupler isn't a mirror.

You may have noticed that a cavity-dumped laser is really just a different type of Q-switched laser, one in which the cavity Q-switches from high to low instead of from low to high. For this reason cavity dumping is sometimes called "pulse transmission mode Q-switching." But "cavity dumping" is easier to say.

What's the duration of the cavity-dumped pulse from the laser in Fig. 12.2? It depends on how long it takes light to make a round trip of the resonator, that is, on the resonator length. The longer the resonator, the longer it takes the pulse to get out. The last photon to emerge is the one that passed through the Pockels cell just before the voltage was

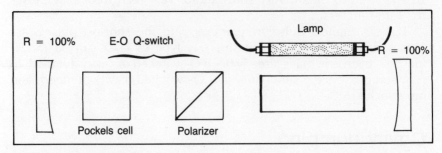

Fig. 12.1   *A cavity-dumped laser with an electro-optic cavity dumper*

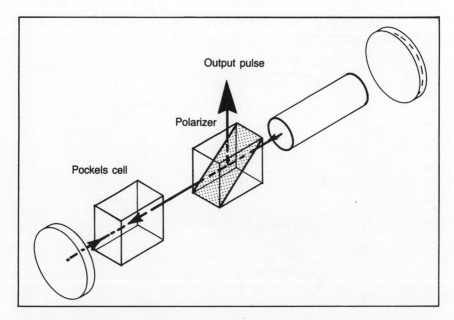

Fig. 12.2 *When the Pockels cell is biased to rotate the light's polarization, a pulse containing all the intracavity energy is dumped out of the laser resonator*

applied, and it must make one round-trip of the resonator before it comes out. The distance it must travel is 2L (L is resonator optical length) and its velocity is c, so the time it takes to do this is 2L/c. Thus, the duration of the cavity-dumped pulse is the time that elapses between the emergence of the very first photon and the very last one, or 2L/c.

This pulse duration can turn out to be very short. If you recall that the speed of light is about 1 ft/ns, you can quickly figure that the cavity-dumped pulse from a foot-long laser is about 2 ns long. That's much shorter than the pulse from most Q-switched lasers.

The difference between a cavity-dumped laser and a Q-switched laser is that the energy is stored in the population inversion in a Q-switched laser and it's stored in the optical resonator (cavity) of a cavity-dumped laser. Of course, you can also store energy in a laser's power supply, as is done in flash-pumped solid-state lasers, in some industrial carbon dioxide lasers, and elsewhere. The flower-pot analogy of the previous chapter is revived in Fig. 12.3, but this time it shows all three places where energy can be stored: the power supply, the population inversion, and the optical resonator. The electrical input

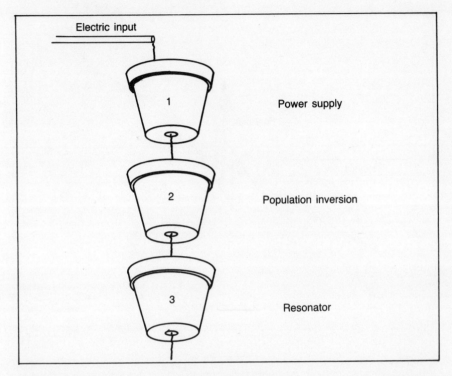

Electric input

Power supply

Population inversion

Resonator

Fig. 12.3 *Like water flowing from one flower pot to another, energy flows from a laser's power supply to its population inversion to its optical resonator. And energy can be stored in any of these places and released later in a high-power pulse*

dribbles into the power supply, and if none of the pots is corked, the output dribbles out of the resonator at the same rate. But if any one (or more) of the pots is corked, energy can be stored in that pot and released in a concentrated, high-power pulse. The three parts of a laser corresponding to the three flower pots of Fig. 12.3 are shown in Fig. 12.4.

A solid-state laser that is flash pumped but not cavity dumped or Q-switched (that is, one with a cork in only the top pot of Fig. 12.3) is known as a *normal mode* laser. The output pulse consists of several spikes and is generally not very repeatable from one pulse to the next. This type of laser is sometimes useful in industrial applications where only crude energy is needed and where the refinement of a Q-switch or cavity dumper would be an unnecessary expense.

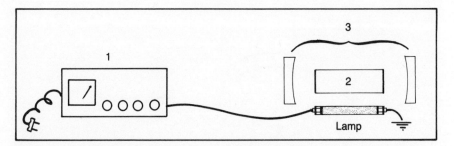

Fig. 12.4   *The flower pots of Fig. 12.3 correspond to a laser's power supply (1), its population inversion (2), and its optical resonator (3)*

Things get more interesting when you start putting corks in more than one of the pots of Fig. 12.3. For example, a flash-pumped, cavity-dumped laser could be represented by putting corks in both the top and bottom pots. The energy would be compressed first in the power supply, and then the pulse would be concentrated further by storage in the optical resonator. The resulting pulse would be more powerful (that is, its energy would be more concentrated) than the pulse produced by either flash pumping or cavity dumping alone. If you corked all three pots, you would have a flash-pumped, Q-switched, cavity-dumped laser.

Partial cavity dumping

When all the energy circulating between the mirrors is dumped out of a laser cavity, it's necessary to wait many microseconds or even milliseconds before the energy can be built up again for a second dump. In partial cavity dumping, only a fraction of the energy between the mirrors is dumped out. In terms of our familiar flower-pot analogy, partial cavity dumping corresponds to keeping the bottom pot half-full and pulling its cork out for only the briefest period so that only a small amount of energy comes out in the pulse. This arrangement is shown in Fig. 12.5.

Of course, energy comes out of the laser on *average* as fast as it goes in. But because the output energy is compressed into pulses, its peak power is greater. This line of reasoning holds for any pulsed laser.

Ion lasers have upper-state lifetimes too short to allow enough energy storage for Q-switching. (The middle flower pot is too leaky.) If you want to obtain a pulsed output from an argon-ion laser or a

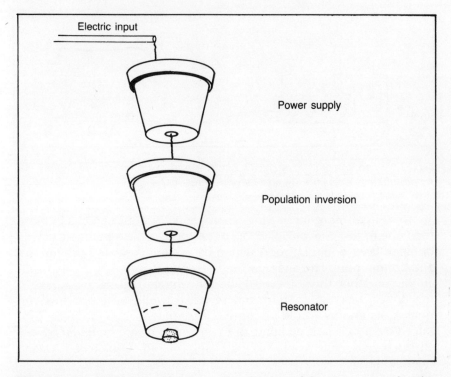

Fig. 12.5 *In partial cavity dumping, only a fraction of the energy stored in the resonator is let out in each pulse*

krypton-ion laser, you must cavity dump it. Such a laser is shown in Fig. 12.6. Here, an acousto-optic cavity dumper ejects part of the intracavity energy to generate the output pulse train. Typically, the signal to the cavity dumper might be an 80-MHz acoustic signal, chopped at 10 kHz as shown in Fig. 12.7. Then the output would be a 10-kHz train of pulses, diffracted from the acousto-optic cavity dumper.[1]

---

[1]*In the previous chapter, we discussed acousto-optic modulators that produce many diffracted beams. These are known as Raman-Nath modulators. Another kind of acousto-optic modulator, called a Bragg modulator, produces only one diffracted beam. Assuming that you want to obtain the output of a cavity-dumped laser in a single beam, you'd use a cavity dumper that operates in the Bragg regime. The difference between the two regimes has to do with the optical and acoustic wavelengths involved and the length of the interaction region.*

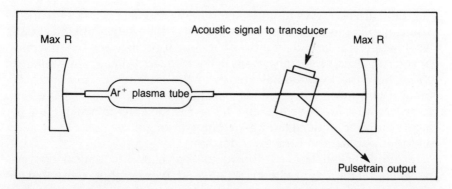

Fig. 12.6 *An argon-ion laser cavity-dumped with an acousto-optic modulator*

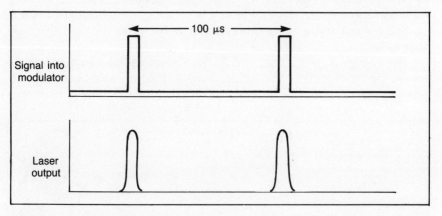

Fig. 12.7 *Acoustic input signal and optical pulsetrain output for partially cavity-dumped laser in Fig. 12.6*

Modelocking—time domain

The shortest pulses of light that have ever been generated have come from modelocked lasers. The duration of a Q-switched laser pulse varies from several hundred nanoseconds to several nanoseconds, depending on the laser parameters. A cavity-dumped pulse can be a little shorter than that—maybe shorter than a nanosecond—but a modelocked pulse from a dye laser can be shorter than a picosecond. That's a thousand times shorter than the 1-ns pulse from a cavity-dumped laser.

There are two ways to understand how a modelocked laser works. You can examine what happens in the time domain by thinking about what happens as laser light moves back and forth between the mirrors, or you can examine what happens in the frequency domain by thinking about how the longitudinal modes of the laser interfere with each other. Either way is correct, and in fact they turn out in the final analysis to be two ways of saying the same thing. But the time-domain picture is a little easier to understand the first time through. We'll discuss it first and then explain the frequency domain.

A modelocked laser is shown in Fig. 12.8. The optical energy between the mirrors has been compressed to a very short pulse that is shorter than the resonator itself. In a Q-switched or a cavity-dumped laser, the whole resonator is filled with energy; but in a modelocked laser the energy is compacted into a pulse that bounces back and forth between the mirrors. Each time this intracavity pulse bounces off the partially transmitting mirror, an output pulse is transmitted through that mirror.

The energy is compacted into the modelocked pulse in the resonator by the modelocking modulator, which is simply a fast optical gate (for example, it could be an electro-optical Q-switch). The gate opens

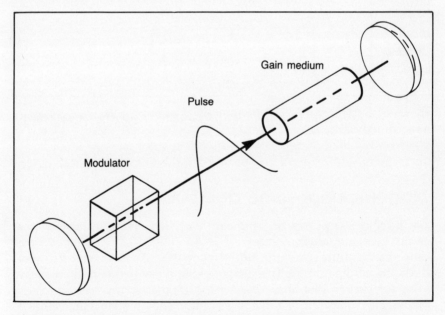

Fig. 12.8 *A short pulse of light bounces back and forth between the mirrors of a modelocked laser*

once per round-trip transit time, letting the pulse through. The rest of the time the gate is closed; the only light that can circulate between the mirrors is the light in the modelocked pulse. The modulator is placed as close as possible to the mirror, and it opens only once while the pulse passes through it, reflects off the mirror, and passes through the modulator again. The average power output from the laser is not affected by modelocking.

The output of a modelocked laser is a train of very short pulses. The time separation between pulses is the distance traveled by the intra-cavity pulse between reflections by the output mirror, divided by its velocity. That is, the period between pulses is 2L/c, where L is again the optical distance between the mirrors. The frequency of modelocked pulses is the reciprocal of the period, or f = c/2L. Thus, a laser whose mirrors are separated by 30 cm will produce a modelocked pulse train at 500 megahertz (MHz)—500 million pulses of light per second.

The duration of the modelocked pulses depends on several factors, including the laser's gain bandwidth and the effectiveness of the mode-locking modulator (its *modulation depth*). The greater the laser's band-width, the shorter the modelocked pulses can be. Thus, dye lasers, which can have enormous bandwidths, produce the shortest mode-locked pulses—sometimes approaching 100 femtoseconds (0.1 ps). On the other hand Nd:YAG lasers, with relatively narrow bandwidths, produce modelocked pulses 30–60 ps in duration.

Electro-optic modulators, acousto-optic modulators, and dye cells have all been used to modelock lasers. An acousto-optic modelocker is shown in Fig. 12.9. Notice that the side of the quartz block opposite the transducer isn't configured to minimize reflection of the incident sound wave as it is in an acousto-optic Q-switch. In fact, an acousto-optic modelocker works differently than a Q-switch. The Q-switch is turned off and on by turning the acoustic signal applied to its trans-ducer off and on. A modelocker reflects the sound wave back across the modulator so that a *standing wave*—like the wave that's formed in a violin string—is produced in the modulator.

Now if you think about a standing wave for a minute, you realize that it "disappears" twice during each cycle. The violin string is at one instant bowed upward, and an instant later it's bowed downward. But between those two extremes, the string is flat; it's momentarily a per-fectly straight line between its two ends. And this momentarily straight string appears twice per cycle between the extremes of the motion.

Likewise, there are two times per period when the elastic medium of a standing-wave acousto-optic modulator is not perturbed by the acoustic wave. During these times light is not diffracted from the mod-

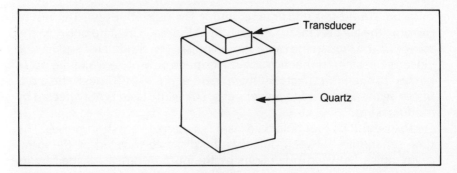

Fig. 12.9 *An acousto-optice modelocker is a standing-wave device in which sound is reflected back toward the transducer from the far side of the modulator*

ulator and a modelocked pulse of light can pass through the modulator without loss.

So if you apply a 100-MHz signal to a standing-wave acousto-optic modulator, there will be 200 million times per second when the modulator doesn't diffract light. If you put the modulator near the mirror of a 75-cm-long laser (c/2L = 200 MHz), it will modelock the laser. The intracavity pulse will pass through the modulator, reflect off the mirror, and pass through the modulator again each of the 200 million times per second that the modulator doesn't diffract light. The output from the laser will be a train of 200 million pulses per second.

Electro-optic modelockers can operate exactly like electro-optic Q-switches, rotating the polarization of light that is subsequently ejected by a polarizer. So you could drive an electro-optic modelocker (composed of a Pockels cells and a polarizer) at 100 MHz and substitute it for the acousto-optic modulator in the previous paragraph.

But there's another way to use an electro-optic modulator to modelock a laser, one that uses phase modulation instead of loss modulation. While a loss modulator—like the electro-optic or acousto-optic modelockers—imposes a loss on light that tries to pass through it when it's closed, a phase modulator imposes a frequency shift on the light. What good does that do? Well, light that tries to get through a phase modulator when it's closed is frequency shifted out from under the center of the laser gain bandwidth. That is, the modulator changes the frequency of the light enough that it can no longer stimulate emission efficiently when it passes through the gain medium. This reduced gain is just as deleterious as attenuation to light that passes through the modulator at the wrong time.

Passive dye cells can also modelock a laser, and they usually Q-switch it at the same time. The output from such a laser would be a pulse train of modelocked pulses having the envelope of a Q-switched pulse. The principle is the same as a passively Q-switched laser: The leading edge of the pulse bleaches the dye, so the rest passes through with minimal loss. Modelocking with a dye cell is something of a hit-or-miss proposition, with typically 50% to 80% of the pulses being well modelocked.

Modelocking—frequency domain

Modelocking is as curious a name as *cavity dumping* is descriptive. What modes are locked, and what does it mean to lock modes? To understand why it's called "modelocking," you have to understand the frequency-domain viewpoint as well as the time-domain viewpoint.

The longitudinal modes of the resonator are locked together in phase when you modelock a laser. (That's why you'll sometimes hear it referred to as *phase locking*.) Suppose you have a laser with three longitudinal modes oscillating simultaneously, as shown schematically in Fig. 12.10. Of course, the waves are moving inside the resonator at about the speed of light. This figure just shows what they look like at one instant in time. Most places in the cavity will be like point A in the figure: The three modes add together to produce a very small total intensity. That is, they interfere with each other destructively. But at one (or maybe more) place in the cavity, all three modes will be at their maximums and they'll add up to a large total. And because all three waves are moving at the speed of light, the spot where they add constructively also moves along at the speed of light. As you've already figured out, constructive interference at this spot is what creates the modelocked pulse.

The problem is that in a *free-running* laser—one that's not modelocked—the three modes of Fig. 12.10 won't stay in phase with each other. Resonator perturbations will cause phase shifting among them, and the pulse will be lost. So it's necessary to lock them together in phase to produce a modelocked pulse. That's what the modelocking modulator does. It transfers energy among all the modes, and this transferred energy contains phase information that prevents the modes from shifting phase with each other.

It turns out that the time-domain viewpoint and the frequency-domain viewpoint are two equivalent ways of looking at the same

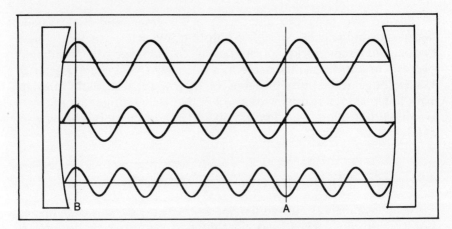

Fig. 12.10   *Three longitudinal modes are shown spatially displaced for clarity. The three modes add constructively at B, destructively at A*

thing, but showing that equivalence requires a level of mathematics (Fourier analysis) beyond the scope of this book.

Applications of modelocked lasers

Modelocked lasers don't usually have very high peak power. The pulses are very short, but there are so many of them that no one pulse can contain much energy. So unamplified modelocked lasers are not found in applications requiring high peak powers. Instead, modelocked lasers are used when very short pulses are needed.

One such application is in communications. Digital information, as you may know, is a collection of ones and zeros—it's the language of computers. Digital information can be sent at very high rates by a modelocked laser. In one simple scheme, a fast modulator placed in front of the laser blocks the modelocked pulse if a zero is to be sent or transmits it if a one is to be sent. Although modelocked lasers are usually not used in the fiberoptic, terrestrial communication links now being integrated into many telephone systems, they are used in experimental systems being developed by the military for communication among spacecraft and with submarines.

Another application of modelocked lasers is in ranging. A modelocked pulse is reflected from a distant object like a satellite, and the time it takes to return to the transmitter is carefully measured. Since the pulse moves at the speed of light, the distance to the object can be readily calculated. Q-switched lasers are also used in ranging. But because their pulses are several meters long, the precision of a Q-switched laser-ranging system cannot be much better than a meter or two. The pulse from a modelocked laser, on the other hand, is only several centimeters long, so ranging with much greater accuracy is possible.

Modelocked lasers are also used as spectroscopic tools to investigate very fast phenomena. Spectroscopy is a whole science by itself, and it's discussed a little in Chapter 16 of this book. Basically, it involves studying matter by observing how light interacts with it. If the matter that you're studying changes very quickly, say, in a fraction of a nanosecond, then the probe you're using to study it must also be that fast. The short pulse from a modelocked laser is one of the few probes that can be used to investigate very fast chemical or physical reactions.

Types of modelocked lasers

Modelocking can be combined with any of the other techniques discussed in this chapter to produce pulsed lasers, or it can be used by itself to produce an unending train of modelocked pulses. For example, a passively modelocked and Q-switched laser is shown in Fig. 12.11. In the top part of this figure, the dye only Q-switches the laser. In the bottom part, the dye concentration has been changed so that it now modelocks and Q-switches the laser.

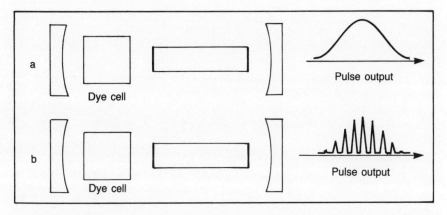

Fig. 12.11 *A dye cell can Q-switch a laser (a) or it can simultaneously Q-switch and modelock the laser (b)*

Questions

1. Sometimes a laser is simultaneously Q-switched and cavity dumped. What would be the advantage of such a laser over one that was only cavity dumped or only Q-switched? How would you design such a laser if you could use only a single electro-optic Q-switch? Sketch the resonator, showing the Q-switch and the mirror transmissions, and sketch a plot of the voltage applied to the Q-switch as a function of time.

2. Suppose a laser were simultaneously cavity dumped and modelocked. What would the output of such a laser look like? Sketch a diagram of the resonator, showing the intracavity devices you might use.

3. Consider a continuouswave, modelocked Nd:YAG laser. If its average power output is 500 mW and the optical distance between its mirrors is 40 cm, calculate its pulse repetition frequency and the peak power of a single modelocked pulse whose duration is 50 ps.

4. A dye cell can simultaneously Q-switch and modelock a laser, as shown in Fig. 12.11. If the optical distance between the mirrors is 30 cm and the duration of the Q-switched pulse curve is 30 ns, how many modelocked pulses are there within the Q-switched pulse?

Chapter

Thirteen

Nonlinear Optics

In the previous three chapters, we discussed modifying the spectral and temporal characteristics of a laser. That is, we talked about how to reduce a laser's spectral width and how to change the temporal shape of its output by several pulsing techniques. In this chapter we'll examine nonlinear optics, the technology that can change the wavelength of light produced by a laser.

Strictly speaking, you don't need a laser to produce nonlinear optical effects, but these effects require such high optical intensities that they are difficult to produce without lasers.

Nonlinear optics is a very useful technology because it extends the usefulness of lasers by increasing the number of wavelengths available. Wavelengths both longer and shorter than the original can be produced by nonlinear optics. There is even one nonlinear device that can convert a fixed-wavelength laser to a continuously tunable one.

This chapter will begin with a discussion of second-harmonic generation, which is probably the single most important type of nonlinear effect. Phase matching, which is absolutely necessary for any efficient nonlinear interaction, will be explained in the context of second-harmonic generation. Then we'll take a look at several other nonlinear effects, including higher harmonic generation and mixing, and parametric oscillation.

What is nonlinear optics?

Nonlinear optics is a completely new effect, unlike anything discussed before in this text. Light of one wavelength is transformed to light of another wavelength—an impressive feat.

This transformation is completely different than, say, a piece of red glass "transforming" white light to red light. The red light was already present in the white light before it hit the piece of red glass. The glass only filters out the other wavelengths; it does not generate a new wavelength. But in nonlinear optics, new wavelengths are generated. A classic example is shown in Fig. 13.1, where the second harmonic—green light at a wavelength of 532 nm—is generated from the 1.06-μm beam of infrared light from a Nd:YAG laser.

Incidentally, it's important to notice that only part of the 1.06-μm light is converted to the second harmonic; part is unchanged. In many cases it is very important to maximize this conversion efficiency, and in the next section we'll see how this can be done.

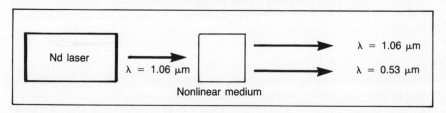

Fig. 13.1 *In second-harmonic generation, part of the light passing through the nonlinear medium is converted to light of one-half the original wavelength*

How is the new wavelength of light created? To gain an intuitive—albeit nonrigorous—understanding of what happens in nonlinear optics, think about the electrons in a nonlinear crystal. (Nonlinear effects can also occur in liquids and gases, but crystals are most common. The explanation we'll give here would also hold, with minor modifications, for a nonlinear liquid or gas.) These electrons are bound in "potential wells," which act very much like tiny springs holding the electrons to lattice points in the crystal, as shown in Fig. 13.2. If an external force pulls an electron away from its equilibrium position, the spring pulls it back with a force proportional to displacement: The spring's restoring force increases *linearly* with the electron's displacement from its equilibrium position.

The electric field in a light wave passing through the crystal exerts a force on the electrons that pulls them away from their equilibrium posi-

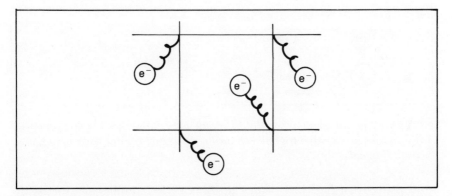

Fig. 13.2 *Electrons in a nonlinear crystal are bound in potential wells, which act something like springs, holding the electrons to lattice points in the crystal*

tions. In an ordinary (i.e., linear) optical material, the electrons oscillate about their equilibrium positions at the frequency of this electronic field. Now, a fundamental law of physics says an oscillating charge will radiate at its frequency of oscillation, so these electrons in the crystal "generate" light at the frequency of the original light wave.

If you think about it for a second, you'll see that this is an intuitive explanation of why light travels more slowly in a crystal—or any dielectric medium—than in a vacuum. Part of the energy in the light wave is converted to motion of the electrons, and this energy is subsequently converted back to light again. But the overall effect is to retard the energy as it moves through the crystal because it takes a detour into the motion of the electrons.

How is a nonlinear material different from the linear material we've been discussing? You can think of a nonlinear material as one whose electrons are bound by very short springs. If the light passing through the material is intense enough, its electric field can pull the electrons so far that they reach the ends of their springs. The restoring force is no longer proportional to displacement; it becomes nonlinear. The electrons are jerked back roughly rather than pulled back smoothly, and they oscillate at frequencies other than the driving frequency of the light wave. These electrons radiate at the new frequencies, generating the new wavelengths of light.

The exact values of the new wavelengths are determined by conservation of energy. The energy of the new photons generated by the nonlinear interaction must be equal to the energy of the photons used. Fig. 13.3 shows the photons involved in the second-harmonic genera-

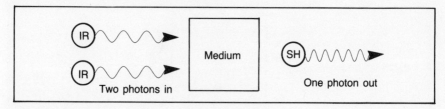

Fig. 13.3 *You can think of second-harmonic generation as a welding process: Two photons are welded together to produce a single photon with the energy of both original photons*

tion process of Fig. 13.1. You can think of the nonlinear process as welding two infrared photons together to produce a single photon of green light. The energy of the two 1.06-μm photons is equal to the energy of the single 532-nm photon.

Another nonlinear process is diagramed in Fig. 13.4. In optical mixing, two photons of differing wavelengths are combined into a single photon of shorter wavelength. What is the new wavelength generated? Recall that photon energy is given by $E = hc/\lambda$. Conservation of energy requires that:

$$\frac{hc}{\lambda_1} + \frac{hc}{\lambda_2} = \frac{hc}{\lambda_3}$$

So the new wavelength is:

$$\lambda_3 = \frac{\lambda_1\lambda_2}{\lambda_1 + \lambda_2}$$

Incidentally, so far we've mentioned two of the three requirements for nonlinear optics: intense light and conservation of energy. The third requirement is conservation of momentum, and that's fulfilled by phase-matching, which will be discussed in a later section of this chapter.

Second-harmonic generation

Second-harmonic generation (SHG), or "frequency doubling," is the most common and probably the most important example of nonlinear optics. It is relatively straightforward compared to other nonlinear interactions, and it can have a relatively high conversion effi-

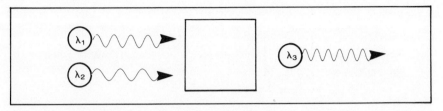

Fig. 13.4 *Optical mixing is similar to second-harmonic generation, except that the original photons have different energies*

ciency. (The conversion efficiency can be defined from Fig. 13.1 as the ratio of second-harmonic power generated to infrared, or fundamental, power input.)

The conversion efficiency of SHG depends on several factors, as summarized in the proportionality:

$$P_{SH} \propto \ell^2 \frac{P_f^2}{A} \left[\frac{\sin^2 \Delta\phi}{(\Delta\phi)^2} \right]$$

where P_{SH} is the second-harmonic power, ℓ is the length of the non-linear crystal, P_f is the fundamental power, A is the cross-sectional area of the beam in the nonlinear crystal, and the quantity in brackets is a phase-match factor that can vary between zero and one. Obviously, it is important to ensure that this factor is as close to unity as possible, but we will postpone a discussion of how this is done until the next section.

Let's take a look at how the factors in the above proportionality affect the harmonic conversion efficiency. For example, Fig. 13.5 shows two identical experiments, except that in the second experiment the nonlinear crystal is twice as long. With the 1-cm crystal, the conversion efficiency is $10^{-9}/10^{-3} = 10^{-6}$. What happens with a 2-cm crystal?.

Contrary to what you might expect, the 2-cm crystal doesn't generate twice as much second harmonic; it generates *four times* as much because the second-harmonic power is proportional to the *square* of the crystal length. That is, 4 nW of 532-nm light is generated in the second experiment. The rest of the light passes through the hypothetical perfect crystal unchanged. (In a real crystal, part of the light would be absorbed and converted to heat.) So in second-harmonic generation, there is such a thing as a free lunch—sort of. With twice as much crystal, you get four times as much second-harmonic output. The drawback is that a 2-cm nonlinear crystal can often cost more than four times as much as a 1-cm crystal.

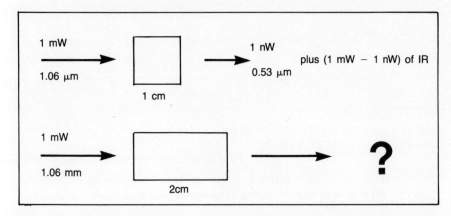

Fig. 13.5 *If the crystal length in the upper experiment is doubled, how much second harmonic is generated?*

In Fig. 13.6 we've gone back to the 1-cm crystal, but now the second experiment has twice as much incoming fundamental power. How much second harmonic is produced in this case?

Since second-harmonic power is proportional to the square of the fundamental power, four times as much second harmonic is generated in the second experiment. As was the case when the crystal length was doubled, a 4 nW of second-harmonic power is generated in the second experiment of Fig. 13.6.

And this fact—that the second-harmonic power generated is proportional to the square of the fundamental power—can be used to advantage with a pulsed laser. The first experiment in Fig. 13.7 shows a 20-ns, 2.5-MW (peak power) pulse of 1.06-μm light that is frequency doubled with 10% efficiency in a nonlinear crystal. In the second experiment the same amount of fundamental energy (2.5 MW × 20 ns = 50 mJ) is compacted into a 10-ns pulse, creating a pulse with twice as much peak power as in the first experiment. How much second-harmonic power is produced in this case?

As before, we've doubled the fundamental power, so the second-harmonic power is quadrupled to 1 MW in Fig. 13.7. Another way to look at this is to solve the proportionality above for conversion efficiency:

$$\frac{P_{SH}}{P_f} \propto l^2 \frac{P_f}{A} \left[\frac{\sin^2 \Delta\phi}{(\Delta\phi)^2} \right]$$

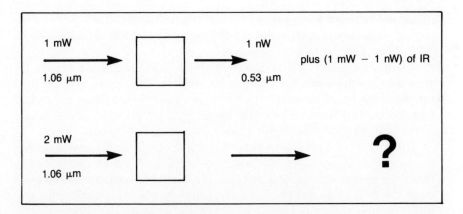

Fig. 13.6 *If the fundamental power in the upper experiment is doubled, how much second harmonic is generated?*

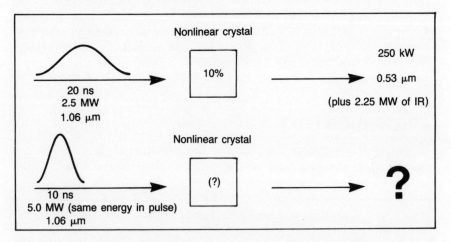

Fig. 13.7 *If the pulse energy in the upper experiment is compacted into a pulse half as long, how much second harmonic is generated?*

Now you can see that conversion efficiency is proportional to the fundamental power. Because the fundamental power was doubled in the second experiment, the conversion efficiency must be 20%. And if 20% of the 5-MW input is frequency doubled, then the second-harmonic power is 20% of 5 MW, or 1 MW.

There's still one more thing you can do to boost the conversion efficiency of second-harmonic generation, and it's shown in Fig. 13.8.

Suppose the beam has a 1-mm radius in the first experiment, and in the second experiment a lens shrinks the beam to a 0.5-mm radius. Now what is the second-harmonic power?

Again, it's 4 nW. Why? Because the second-harmonic power is inversely proportional to beam area, which is proportional to the square of the beam radius ($A = \pi r^2$). So by reducing the beam radius by a factor of two, you reduce the beam area by a factor of four and increase the second-harmonic power by a factor of four.

These are precisely the kinds of things you do when you want to increase the conversion efficiency of second-harmonic generation: You get a longer crystal, you increase the fundamental power, and you focus the beam into the crystal. Unfortunately, there are limitations on all these tricks. If you focus too tightly or increase the fundamental power too much, you may damage the expensive nonlinear crystal.

And there is another limitation on focusing. Remember from Chapter 9 that a tightly focused Gaussian beam has greater divergence than one that isn't focused to such a small spot. It turns out that a diverging beam is less efficient at second-harmonic generation than a collimated one. (It has to do with phase-matching.) So even if you don't damage the crystal, you still can't focus as tightly as you might like to into a nonlinear crystal.

Phase-matching

None of the things discussed in the previous section makes any difference if the last term of the proportionality—the phase-match

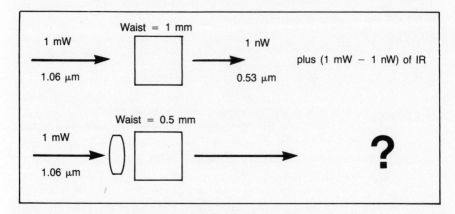

Fig. 13.8 *If the beam in the upper experiment is focused to one-half its original diameter, how much second harmonic is generated?*

term—is equal to zero. Phase-matching is vital to any nonlinear inter-action, but we'll discuss it here in the context of second-harmonic gen-eration. A similar concept holds for any other nonlinear interaction.

First, let's understand the problem; then we'll look for a solution. Fig. 13.9 shows the problem. If the second-harmonic power generated at point B is out of phase with the second-harmonic power generated at point A, they will interfere destructively and result in a total of zero second harmonic from the two points. What's worse, if the crystal isn't phase-matched, the second harmonic generated at nearly every point in the crystal will be canceled by a second harmonic from another point. Practically no second harmonic will be produced, no matter how tightly you focus or how long the crystal is.

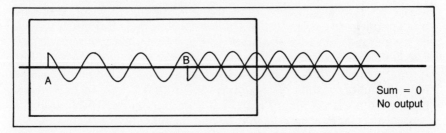

Fig. 13.9 *If a nonlinear crystal is not phase-matched, harmonic light generated at one point will interfere destructively with that generated at another point*

How does the second harmonic get out of phase with itself in the first place? That is, the second harmonic from point B is generated from the same fundamental wave that generated the second harmonic at point A. Obviously, the fundamental wave stays in phase with itself as it propagates through the crystal. Why isn't all the second harmonic auto-matically generated in phase with itself?

Dispersion is the answer to that question. Remember that the refractive index of the nonlinear crystal is slightly different for the two wavelengths. Therefore, although the second-harmonic wavelength is exactly half as long as the fundamental wavelength in vacuum, that is not true inside the crystal. The frequency of the second harmonic is still exactly twice that of the fundamental, but the wavelength relationship between the two is altered by dispersion. If you're confused at this point, it might be helpful to go back to and review the text relating to Figs. 3.3 and 3.4.

So dispersion is the problem. It causes the phase between the fun-damental and the second harmonic to shift slightly as the two travel along together inside the nonlinear crystal. Eventually, the phase shift becomes large enough so that new second-harmonic light is generated exactly 180° out of phase with the original second harmonic.

In many cases the harmonic is generated in a polarization orthogonal to the fundamental. Then the problem can be solved by using a birefringent crystal as the nonlinear medium. Remember that in a birefringent crystal the refractive index depends on the polarization of light. So if you choose a birefringent material whose ordinary index at the fundamental wavelength is equal to its extraordinary index at the harmonic wavelength, you can avoid the phase-matching problems caused by dispersion. Because the refractive index is the same for both wavelengths, there is no dispersion.

Of course, it's not easy to find nonlinear crystals whose refractive indices satisfy these conditions. In fact, in every case you have to resort to some sort of gimmick to phase-match a nonlinear crystal. There are two gimmicks in common use: temperature tuning and angle tuning. In temperature tuning, you start with a crystal whose indices almost satisfy your requirements and you heat or cool it until the (temperature-dependent) indices are exactly right. In angle tuning, you change the orientation of the crystal with respect to the incoming laser beam. Because the extraordinary refractive index depends on the angle of propagation, it can be adjusted in this manner until its value for the harmonic is equal to the ordinary index's value for the fundamental.

The foregoing discussion of phase-matching has applied only to one type of phase-matching; there is another type that is more difficult to explain intuitively. What's been explained here is Type I phase-matching, in which the fundamental light is in the ordinary polarization of the nonlinear crystal and the second-harmonic is generated in the extraordinary polarization. In Type II phase-matching, the fundamental is evenly divided between the ordinary and extraordinary polarizations and the second harmonic is generated in the extraordinary polarization. There is no nonmathematical explanation of Type II phase-matching corresponding to the discussion here of Type I. But it turns out that in some cases, such as SHG with very high-power, solid-state lasers, Type II phase-matching is more efficient than Type I. The orientation of the electric fields for Type I and Type II phase-matching are shown in Fig. 13.10.

Some of the nonlinear materials that can be phase-matched for SHG of 1-μm laser light (and for other nonlinear interactions) are shown in the table on page 175.

There are several things to notice in this table. First, the materials that have high nonlinearity are easily damaged, so they are used only with low-power lasers. Several of the crystals are available in both deu-

Crystal type	Relative nonlinearity	Long pulse damage threshold MW/cm^2
KDP (potassium dihydrogen phosphate)	1.0	400
KD*P (deuterated KDP)	1.06	400
CDA (cesium dihydrogen arsenide)	0.92	300
CD*A (deuterated CDA)	0.92	300
ADP (ammonium dihydrogen phosphate)	1.2	400
LiNbO$_3$ (lithium niobate)	13.4	6–40
BSN (barium sodium niobate)	38.0	10–25

terated and undeuterated forms. In the deuterated case, many of the hydrogen atoms in the crystal are replaced with the hydrogen isotope deuterium. This replacement often enhances nonlinearity and reduces infrared absorption, but it can degrade the crystal's optical quality. Long pulse damage thresholds are those observed for nanosecond pulses from Q-switched lasers. Different thresholds are often observed for picosecond pulses from modelocked lasers. Finally, the damage thresholds vary quite a bit from crystal to crystal, so the values given here are rough averages.

Higher harmonics

Third-harmonic light can be generated with an arrangement very similar to second-harmonic generation, as shown in Fig. 13.11a. But phase-matching requirements make it very difficult to generate the third harmonic in a single step in a crystal, so a two-step process is common. As shown in Fig. 13.11b, the second harmonic is generated in the first crystal and is then "mixed" with the fundamental in the second crystal to produce the third harmonic.

Fourth, fifth, and higher harmonics can also be generated, but the efficiency of these processes is generally quite low. Even higher harmonics of lasers have been generated experimentally, but the purpose

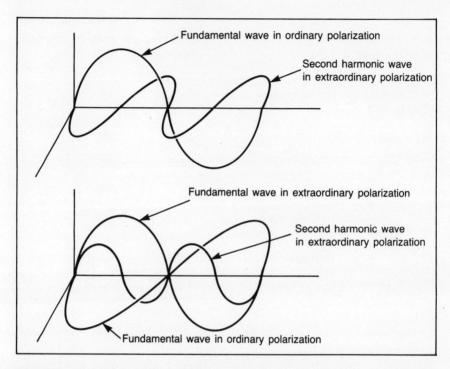

Fig. 13.10 *For Type I phase-matching (top), fundamental light is in ordinary polarization and second harmonic is in extraordinary polarization. For Type II phase-matching (bottom), fundamental light is polarized midway between ordinary and extraordinary directions, and second harmonic is polarized as extraordinary*

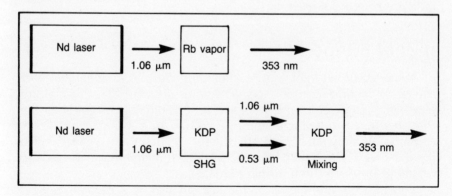

Fig. 13.11 *(a) Single-step, third-harmonic generation. (b) Generation of third harmonic by SHG and mixing*

of these experiments was more to investigate the properties of the nonlinear media than to generate useful amounts of short-wavelength light.

Optical parametric oscillation

So far, all the nonlinear interactions we've discussed involve combining the energy of one or more photons into a single, more-energetic (shorter-wavelength) photon. But the process can also work the other way: The energy in one photon can be divided among two new photons. That's what happens in an optical parametric oscillator (OPO), as shown in Fig. 13.12.

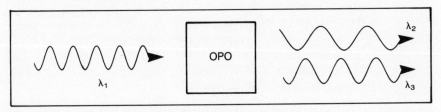

Fig. 13.12 *An OPO generates two wavelengths from a single input wavelength*

Conservation of energy must hold among the photons involved, that is:

$$\frac{hc}{\lambda_1} = \frac{hc}{\lambda_2} + \frac{hc}{\lambda_3}$$

An OPO is an oscillator. Unlike the other examples of nonlinear optics we've discussed, an OPO must have mirrors like a laser to form an optical resonator. (But what happens in an OPO is only nonlinear optics; there is no stimulated emission.) Fig. 13.13 shows a singly resonant OPO in which only one wavelength—called the idler—is reflected by the mirrors. In a doubly resonant OPO, both the pump (λ_1) and the idler (λ_3) are reflected, and only the signal (λ_2) is transmitted.

What determines the output wavelength (λ_2) of the OPO in Fig. 13.14? The equation above can be solved to yield:

$$\lambda_2 = \frac{\lambda_1\lambda_3}{\lambda_3 - \lambda_1}$$

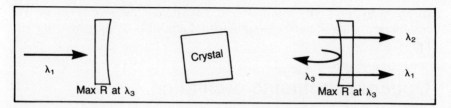

Fig. 13.13 *In a singly resonant OPO, only one wavelength is reflected from the mirrors*

But λ_2 is not uniquely determined by this equation because λ_3 can have any value. Does this mean than an OPO generates light of many wavelengths?

The answer is yes, but only one wavelength at a time because only one wavelength is phase-matched at a time. It is this capability to generate many wavelengths that makes an OPO important: You can tune an OPO to generate the wavelength you want. If you want light whose wavelength is 1.48 μm, for example, you could generate it with an OPO that was pumped by a Nd:YAG laser.

The phase-matched wavelengths of an OPO are changed by adjusting the nonlinear crystal's temperature or angle. Fig. 13.14 shows the tuning curve of an angle-tuned, Nd:YAG-pumped LiNbO$_3$ OPO. If the propagation angle (with respect to the crystal's optic axis) is 46°, then the signal and idler wavelengths will be about 1.6 μm and 3.1 μm, respectively. (It's arbitrary which one is called the "signal" and which is called the "idler.")

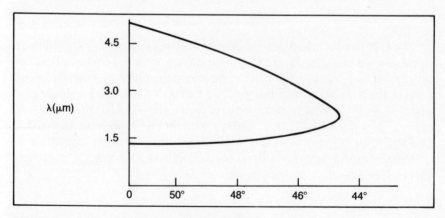

Fig. 13.14 *Tuning curve for a LiNbO$_3$ OPO*

Questions

1. What is the wavelength of the second harmonic of a chromium-ruby laser? What is the third-harmonic wavelength of a Nd:YAG laser?

2. Suppose the second crystal in Fig. 13.5 were 3 cm long instead of 2 cm. How much second harmonic would be generated?

3. The modelocked laser shown below produces 1 W of light at its fundamental wavelength, and 10 mW of second harmonic is normally generated by the nonlinear crystal. But if the power to the acousto-optic modelocking modulator is increased, 15 mW of second harmonic can be produced. Calculate the fractional change in pulse duration caused by increasing the power to the modulator.

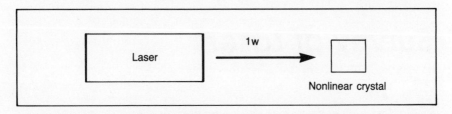

4. Can a 750-nm output (signal) be obtained from a Nd:YAG-pumped OPO? Why? Suppose the Nd:YAG laser is frequency-doubled to produce 530 nm of light. Can this light be used to generate a 750-nm signal from the OPO? What is the idler wavelength in this case?

5. An OPO is degenerate when its signal and idler are the same wavelength. What is this wavelength for a normal Nd:YAG-pumped OPO? For an OPO pumped by a frequency-doubled Nd:YAG?

Chapter

Fourteen

Survey of Lasers

So far, we've discussed the principles of lasers, but we haven't spent much time on specific lasers. In this chapter we'll look in some detail at several different types of lasers, seeing how the principles we've described in previous chapters apply in specific cases.

The most ubiquitous of lasers is surely the helium-neon (HeNe) laser. Its bright red beam finds application in thousands of supermarket checkout stands, in home videodisk players, in laser printers, and in hundreds of different alignment tasks. We'll take a look at the operation of this laser and see which of its parameters are important in practical applications.

We'll also examine molecular lasers, including the carbon-dioxide laser, which finds widespread industrial applications. Solid-state lasers comprise the third important class we'll discuss, and we'll conclude this chapter with a review of organic dye lasers and ion lasers.

Helium-neon lasers

The HeNe laser is is an electrically pumped gas laser that usually produces a red beam at 632.8 nm, but there are also infrared lines at 1.15 and 3.39 μm. Normally, the laser is designed so these infrared lines don't oscillate. Fig. 14.1 is a schematic of a HeNe laser tube. A high-voltage (kilovolts), low-current (milliamps) discharge between the can-shaped cathode and the small anode provides the energy for the population inversion. The discharge is confined to a small inner tube, called the "bore," to keep it concentrated for maximum energy transfer to the tiny beam of circulating power bouncing back and forth between the mirrors. Even so, the electrical efficiency of this laser is typically low—usually less than 0.1%.

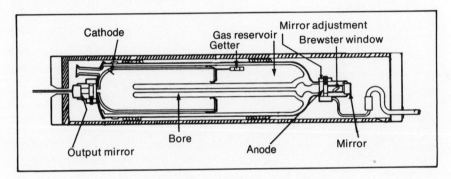

Fig. 14.1 *Schematic of HeNe laser tube*

Energy is transferred from the current when electrons collide with gas atoms in the tube. When a fast-moving electron runs into a ground-state atom, part of the electron's kinetic energy can be transferred to the atom, leaving it in an excited state. But creating a population inversion in a HeNe laser isn't as straightforward as that. It turns out that neon, the atom that lases, is very poor at absorbing energy from the discharge current. Neon atoms appear to have a very small cross section to the electrons, so few electrons collide with and transfer energy to the neon atoms. Helium atoms, on the other hand, seem to have a large cross section to the electrons, and energy is easily transferred from the current to them. Fortunately, the excited helium energy levels are quite close to the upper laser level(s) in neon, so energy can readily be transferred from an excited helium atom to a ground-state neon atom when the two collide.

An energy-level diagram for both helium and neon is shown in Fig. 14.2. After a helium atom is excited to one of its upper states by collision with an electron, it can collide with a neon atom and transfer its energy to the neon, leaving the neon atom in its upper laser level. Exactly which of the three transitions lases depends on the feedback of the resonator; usually visible output is desired, so laser mirrors with

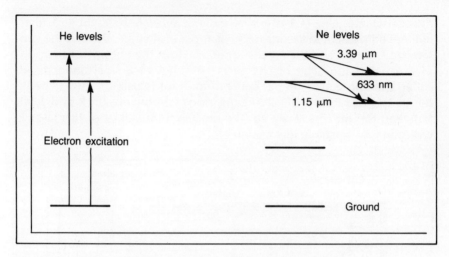

Fig. 14.2 *Helium and neon energy levels*

low infrared reflectivity are used. From its lower laser level, the neon atom decays spontaneously to its ground level.

Output power from commercial HeNe lasers varies from a fraction of a milliwatt up to about 50 mW. In some of the higher-power lasers, residual mirror reflectivity in the infrared cannot be kept low enough to suppress the 3.39-μm transition. (That is, despite low-reflectivity mirrors, round-trip gain exceeds round-trip loss, and lasing occurs at 3.39 μm.) In these lasers the tube is lined with powerful magnets that weaken the 3.39 μm by an effect known as *Zeeman splitting* so that only the 632.8-nm transition is above threshold.

Most modern HeNe lasers use internal mirrors, mirrors that are attached directly to the ends of the discharge tube. Older, and some special-purpose, lasers have external mirrors that are separate from the tube. In these lasers a nonreflecting window, often at Brewster's angle (see Chapter 3), must be placed on the end of the discharge tube. Because the internal-mirror design has greater reliability, it is almost always preferred. The laser is aligned at the factory and usually requires no further adjustment.

The technology of attaching the mirrors to the tube is fairly sophisticated. For many years the mirrors were epoxied on, but epoxy seals are unavoidably porous and allow water vapor to enter the tube over a period of years. As a result, the shelf life of epoxied HeNe lasers was limited to several years. Nearly all commercial HeNe lasers now have weld-like "hard" seals between the mirrors and discharge tube, and their shelf lives are at least a decade. The ultimate limitation on shelf life now appears to be diffusion of helium through the walls of the tube.

Operating lifetime of most commercial HeNe lasers is about 15,000 or 20,000 hours, and the most common failure mechanism is cathode decay. In fact, there's a direct tradeoff between the length of the bore and the tube's lifetime. In a tube with a long bore, the intracavity beam and current can interact over a long distance, and output power is maximized. But if the bore extends almost all the way to the other end of the tube, only a small area of the cathode emits electrons, and the cathode fails relatively quickly. In a short-bore tube, on the other hand, a large area of the cathode can be used to ensure long lifetime, but tube power is decreased.

Molecular lasers

Most lasers rely on population inversions between electronic energy levels in atoms for the stimulated emission that creates laser light. But as we discussed in Chapter 6, molecules also have energy levels. And there are lasers that operate on the rotational, vibrational, and electronic energy levels of molecules. By far the most important of these—from an economic standpoint—is the carbon dioxide (CO_2) laser.

The CO_2 laser is based on vibrational transitions of the CO_2 molecules, so it's important to understand the vibrational modes of the molecule if you want to understand how the laser works. Fig. 14.3 is a diagram of a CO_2 molecule in which the electromagnetic forces holding the molecule together are represented as tiny springs between the atoms.

There are three different, independent modes of vibration possible for the molecule, as diagramed in Fig. 14.3. In the *symmetric stretch mode,* the carbon atom is motionless while the two oxygen molecules vibrate in and out in phase with each other. Notice that the molecule's center of mass doesn't vibrate because the oxygen atoms move out and then back in at the same time. The center of mass must not vibrate in any vibrational mode of any molecule.

In the *bending mode* the molecule flexes so that the two oxygen atoms are twisted upward while the carbon atom twists downward. The motion is such that the center of mass does not vibrate.

In the *asymmetric stretch* mode, the atoms stretch asymmetrically (see Fig. 14.3). But the displacements of the three atoms are such that the center of mass of the molecule does not vibrate.

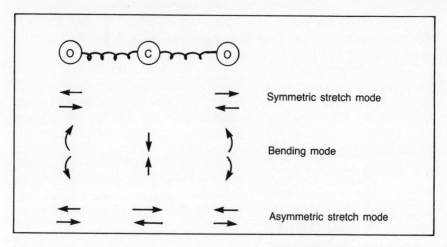

Fig. 14.3   *A carbon dioxide molecule and its vibrational modes*

It's important to remember that the energy in each of these vibrational modes is quantized—that is, only certain amounts of energy are possible in each mode. And the amounts are different in each of the three modes. Fig. 14.4 is a graphical representation of these facts, showing the energy levels of all three modes in a single energy-level diagram. The quantum numbers adjacent to each level tell you which modes are excited. For example, the (1,0,0) mode is the first excited level of the symmetric-stretch mode. The second excited level of the bending mode is designated (0,2,0), and the first excited state of the asymmetric-stretch mode is (0,0,1). Of course, the ground level is (0,0,0).

Only the four lowest excited levels are shown in Fig. 14.4 because they're the only ones that take part in the lasing process. But higher levels exist, including levels in which the molecule is vibrating in more than one mode. For example, a CO_2 molecule in the (1,5,0) mode is flexing (bending) back and forth fairly rapidly while simultaneously stretching symmetrically fairly slowly.

What happens when carbon dioxide lases? First, you have to get some energy into the upper laser level (0,0,1); this is accomplished by collisional excitation. Usually a nitrogen molecule absorbs energy by colliding with an electron and transfers its energy to the CO_2 molecule, just as the helium atom transfers its energy in a HeNe laser. Once in the upper laser level, the CO_2 molecule can be stimualted to emit a 10.6-μm photon and decay to the (1,0,0) lower laser level, or it can be stimulated to emit a 9.6-μm photon and decay to the (0,2,0) level. In

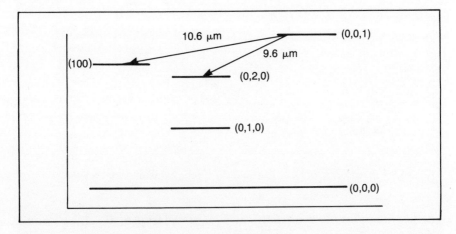

Fig. 14.4 *Vibration energy levels of carbon dioxide*

either case the molecule subsequently decays spontaneously to the (0,0,0) level. In a real CO_2 laser, the output wavelength is selected by providing (resonator) feedback only at the desired wavelength.

Most CO_2 lasers require a mixture of three gases to operate: carbon dioxide, nitrogen, and helium. As we've seen, the nitrogen provides collisional excitation for the carbon dioxide. Helium has high thermal conductivity, and it's in the mixture to help keep the gases cool. The helium provides an efficient vehicle for transporting heat to the outside of the laser.

In addition to vibrational energy, the CO_2 molecule can also store energy in electronic and rotational states. What are the effects of these energies on the diagram in Fig. 14.4?

All the levels shown in Fig. 14.4 are vibrational levels of the ground electronic state. There is a similar set of vibrational levels in each of the higher electronic states, although the exact vibrational energies of these levels are displaced slightly. Lasing has been observed on these so-called "sequence" bands of carbon dioxide, but such lasers are rare and of little economic importance. All commercial CO_2 lasers are based on vibrational transitions within the ground electronic state.

The rotational levels are more important to the real-world operation of CO_2 lasers. The energy-level diagram in Fig. 14.4 is actually a simplification because it doesn't show the rotational structure of the vibrational levels. Fig. 14.5 represents the (0,0,1) and (0,2,0) levels in more detail, showing the many rotational sublevels in each of the vibrational levels. When the molecule undergoes a transition from (0,0,1) to

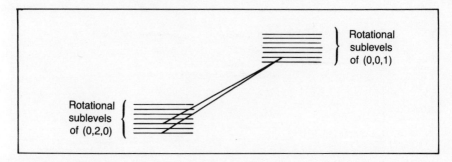

Fig. 14.5   *Rotation sublevels of the 0,0,1 and 0,2,0 vibrational levels*

(0,2,0), it begins from one particular rotational level of (0,0,1) and winds up on one particular level of (0,2,0).

But the molecule can't go to just any rotational level of (0,2,0). A selection rule (based on the conservation laws of physics) requires the rotational quantum number must change by one during the transition. That is, if the molecule starts on the seventh rotational level of (0,0,1), it must wind up on either the sixth or eighth rotational level of (0,2,0).

Knowing this, you can understand the spectral features of a CO_2 laser as shown in Fig. 14.6. There are two branches, each with many lines. The R branch consists of all the transitions in which the rotational quantum number decreases by one, that is, from the first rotational level of (0,0,1) to the zero rotational level of (0,2,0), and so on. And the P branch consists of all the transitions in which the rotational quantum number increases by one.

Carbon dioxide lasers are commercially important because they can generate very high average powers with relatively high electrical efficiency. Commercial CO_2 lasers with average power greater than 20 kW are available, and kilowatt CO_2 lasers are common. Their electrical efficiency can exceed 10%, making them the most energy-efficient lasers in existence. These powerful lasers find a variety of applications in cutting, welding, and heat-treating materials. CO_2 lasers are also important in medical and research applications.

A newer type of molecular laser is making inroads in several applications. In an *excimer* laser, a molecule is formed from atoms that don't normally combine to make molecules. For example, a rare gas such as argon and a halogen such as fluorine can combine in an excimer molecule of argon fluoride.

How is this possible? Any high school chemistry student knows that the rare gases are inert; they don't form molecules.

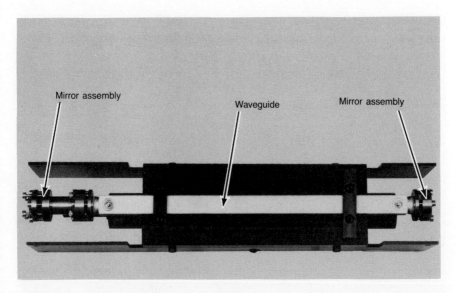

Plate 14.1 *This waveguide carbon-dioxide laser is about a foot long, and produces about 10 watts* (Courtesy California Laser)

It's true that the rare gases don't form stable molecules in the ground state. But they can form stable molecules in excited states. The situation is shown in Figs. 14.7a and b. If both atoms are in the ground state and there's no extra energy around anywhere, then the atoms repel each other as shown in Fig. 14.7a. The closer together the atoms come, the harder they push each other away. They will never combine to form a molecule.

But if there is some extra energy available, perhaps in the form of a fast-moving electron, then a molecule can be created, as shown in Fig. 14.7b. The new molecule is in an (electronic) excited state and is

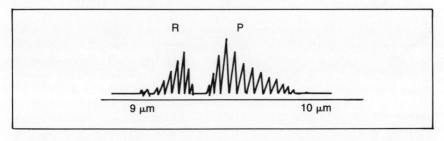

Fig. 14.6 *The R and P branches of the 0,0,1 to 0,2,0 transition in carbon dioxide*

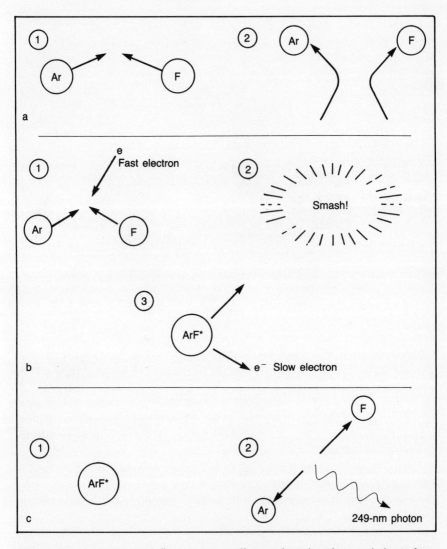

Fig. 14.7   *(a) Argon and fluorine normally repel each other and don't form molecules, but if some extra energy is available (b), a stable ArF molecule can be formed in an excited state. (c) When the excited ArF molecule decays to its ground state with the emission of a photon, it comes apart into an argon atom and a fluorine atom*

stable only as long as it stays in that state. Eventually, its spontaneous lifetime will run out, and the molecule will decay to its repulsive ground state with the emission of a photon, as shown in Fig. 14.7c.

But it is possible to create a population inversion in a collection of argon and fluorine atoms, so there are more molecules than free argon atoms. Then you can have stimulated emission, just as you can from any other population inversion. This is precisely what is done in an excimer laser. The most common rare-gas-halide lasers and their wavelengths are:

Argon fluoride	(ArF)	193nm
Krypton fluoride	(KrF)	249 nm
Xenon chloride	(XeCl)	308 nm
Xenon fluoride	(XeF)	350 nm

From this list, you can deduce why excimer lasers are important: They're a good source of intense ultraviolet light. Before the technology of excimer lasers was developed, ultraviolet lasers could be simulated by combining nonlinear optics with visible-wavelength lasers, but this approach was clumsy.

There are also rare-gas dimer excimer lasers, and rare-gas oxide excimer lasers:

Ar_2	125 nm
Kr_2	146 nm
KrO	558 nm

Solid-state lasers

In solid-state lasers the lasing atoms are embedded in a solid piece of transparent material called the *host*. The most common examples are neodymium embedded in yttrium aluminum garnet crystals (Nd:YAG lasers), neodymium in glass (Nd:glass lasers), and chromium in ruby crystals (Cr:ruby). Neodymium lases at 1.06 μm and chromium in ruby lases at 694.3 nm.

The host material is usually in the shape of a rod roughly the size of a short pencil. Optical pumping provides the energy to create a population inversion. In some solid-state lasers the pump lamp is wrapped helically around the laser rod, but a more common pumping geometry is shown in Fig. 14.8. In this end view of the laser, the cylindrical lamp and laser rod are placed close to each other, each at a focus line of the elliptical pump cavity. The focusing characteristics of the ellipse ensure that any light ray emitted from the lamp will be reflected back to the laser rod, as shown in the figure.

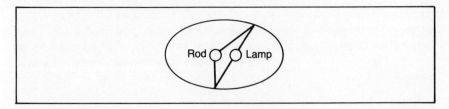

Fig. 14.8 *An elliptical pump cavity for a solid-state laser*

A simplified energy-level diagram from Cr:ruby is shown in Fig. 14.9a and for Nd:YAG or Nd:glass in Fig. 14.9b. Notice that the Cr:ruby laser is a three-level system, while the Nd laser is a four-level system. Unlike the HeNe or CO_2 laser, the active ion in most solid-state lasers absorbs the energy directly without having it go through a "middleman" absorber.

Solid-state lasers are usually water cooled, and often the water is channeled through a glass flow tube around the laser rod. As a result, significant thermal gradients can exist in the rod. These thermal gradients induce a mechanical stress that is detrimental to the laser for several reasons. Stress-induced inhomogeneities in the rod's refractive index can cause birefringence, which introduces a significant intracavity loss mechanism in a polarized laser. As a rule of thumb, thermal birefringence will reduce the polarized output power from a continuous-wave, solid-state laser by about a factor of two.

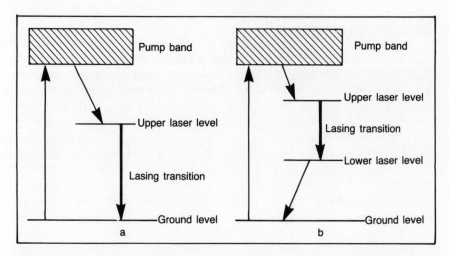

Fig. 14.9 *Energy level diagram for (a) Cr:ruby and (b) Nd in YAG or glass*

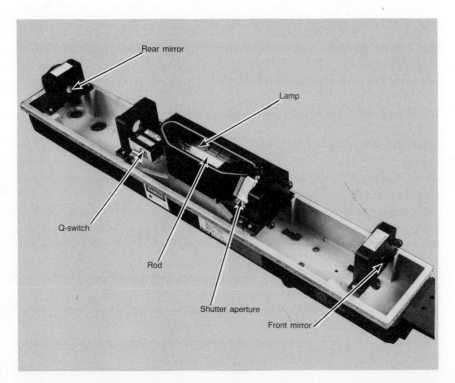

Plate 14.2 *A Q-switched Nd:YAG laser* (Courtesy Electro Scientific Industries)

Thermally induced stress in the laser rod also causes the rod to act as a weak lens, focusing light passing through it. To the extent that this is a uniform effect, it can be compensated by reducing the focusing effect of the laser mirrors. But the degree of focusing depends on location within the laser rod, polarization of light, and pump power. So in general this effect cannot be completely corrected, and it causes a further reduction in output power.

By far the largest application of solid-state lasers is in tactical military equipment, where they are used in rangefinders and target designators. Beyond that, solid-state lasers are important in materials processing, in medicine, and in scientific research.

Organic dye lasers

In Chapter 13 we discussed the parametric oscillator, a nonlinear device that can produce wavelength-tunable laser light. Dye lasers provide an alternative approach to wavelength tunability. The active medium in a dye laser is an organic dye molecule, usually dissolved in a liquid solvent. The dye molecule is very large with many rotational and vibrational levels associated with each electronic level. These rotational and vibrational levels tend to be so numerous that they blend together, producing very broad electronic states. When the molecule drops from one broad electronic state to another, the wavelength of light emitted depends on the starting and ending point within these states. As a result, the emission bandwidth of the dye can be quite wide—as great as 100 nm in some dyes, as illustrated in Fig. 14.10.

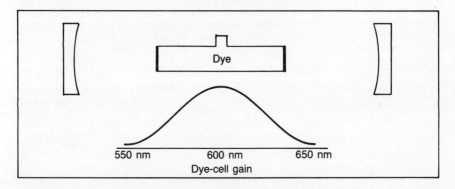

Fig. 14.10 *The gain bandwidth of a dye laser is very large, and the output wavelength can be tuned across the entire bandwidth*

The lasing bandwidth of a dye laser is reduced by limiting the bandwidth of feedback provided by the resonator, as discussed in detail in Chapter 10. Prisms, gratings, birefringent filters, and other devices can be used to reduce the bandwidth of dye lasers. But here is the important point: Because the natural bandwidth of the dye is so great, the laser can be tuned across this whole region by adjusting the feedback wavelength of the resonator. If one or more prisms were introduced into the laser shown in Fig. 14.10, its narrowband output could be tuned most of the way across its 100-nm natural bandwidth.

Dye lasers are almost always optically pumped, either by a flashlamp or by another laser, as shown in Fig. 14.11. As is the case for any optically pumped laser, the pump wavelength must be shorter than the output wavelength. (Do you know why?) Therefore, the tunable output of the dye laser in Fig. 14.11 always has a longer wavelength than the pump laser. If you want a dye laser to produce wavelength-tunable blue light, you must pump it with an ultraviolet laser.

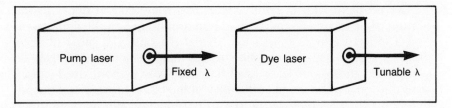

Fig. 14.11 *Dye lasers are often optically pumped by other lasers*

Typical energy levels of a dye laser are diagramed in Fig. 14.12. Notice that there are two types of electronic states: singlets and triplets. (These names derive from the way the total spin of the electrons adds up, but it's not necessary to understand this detail of quantum mechanics to understand how the dye laser works.) Instead being sharp and narrow, these levels are broad and fuzzy because they are composed of many vibrational and rotational sublevels.

The arrows in Fig. 14.12 show the way normal lasing works in an organic dye. In thermal equilibrium, the molecule starts near the bottom of the ground singlet state, S_0. It absorbs a pump photon and is boosted to somewhere near the top of the first excited singlet, S_1. The molecule will lose some energy as it decays to a lower vibrational level within S_1 until it's stimulated to lase and drop back down to S_0. The exact wavelength emitted will depend on where the transition began in S_1 and where it ended in S_0.

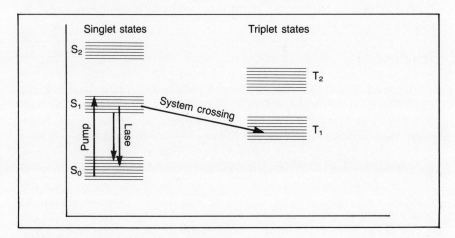

Fig. 14.12 *Energy levels in a dye laser*

The problem with organic dyes is that they tend to decay to the bottom triplet state (T_1) rather than wait around in S_1 to be stimulated. And a molecule in T_1 is bad for several reasons. The most obvious disadvantage of a *system crossing* to T_1 is that the molecule that's been pumped to the upper laser level has been lost without producing a laser photon. So the overall efficiency of the lasing process is reduced. If very many molecules decay to T_1, the reduction can be drastic.

But there are more problems. Once a molecule is in T_1, it tends to stay there; the spontaneous lifetime of T_1 is fairly long. Now, notice that there is a second triplet state, T_2, just above T_1. In many dye molecules, the spacing between T_1 and T_2 is pretty close to the spacing between S_0 and S_1. That means that a molecule in T_1 can absorb a laser photon emitted when another molecule drops from S_1 to S_0. So a population of molecules in T_1 is a built-in loss mechanism for laser photons.

Another problem with the T_1 level is that molecules in this level are less stable chemically, so producing a population in T_1 tends to reduce dye lifetime.

These last two problems can be alleviated by introducing a quenching agent that absorbs energy from the dye's T_1 level. Most modern dye lasers include such a quenching agent. System crossings are still a problem with dye lasers, but they are far less important than they would be without the quenching agents.

Dye lasers are used almost exclusively in research applications where their tunable output is a powerful tool for spectroscopy and photochemistry. Dye lasers could find an important economic application if laser enrichment of isotopes, particularly uranium isotopes, proves feasible.

Ion lasers

Ion lasers are the highest-power continuous-wave visible lasers available, with output as high as 20 W commercially available. There are two common types of ion lasers: argon lasers and krypton lasers. Many different wavelengths are available from each of these lasers, as shown in Fig. 14.13 and Fig. 14.14. Each of these output lines results from its own transition within the gain medium. The most common argon wavelengths are the blue line at 488 nm and the green line at 514.5 nm. Krypton's strongest output is the red line at 647.1 nm, but it can also provide yellow, green, and blue light.

If an ion laser is not restricted to oscillate on a single transition, it will produce several different colors in its output, as shown in

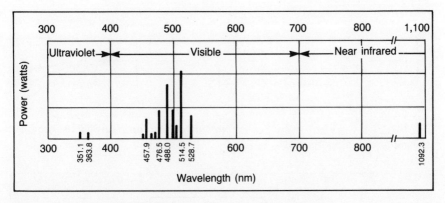

Fig. 14.13 *Wavelengths available from an argon-ion laser*

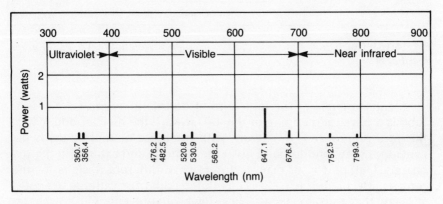

Fig. 14.14 *Wavelengths available from a krypton-ion laser*

Fig. 14.15. An intracavity prism can eliminate the feedback for all but one of these lines, resulting in single-wavelength operation, as shown in Fig. 14.16.

Unlike other gas lasers which operate with low-current, high-voltage discharges, ion laser discharges are relatively low voltage (several hundred volts) and high current (several tens of amperes). The high current is necessary to ionize the argon or krypton gas in the plasma tube, and current density in the tube is typically several hundred or one thousand amps per square centimeter. This high-current density and the resulting power dissipation require careful electrode design and water cooling of all but the lowest-power ion lasers.

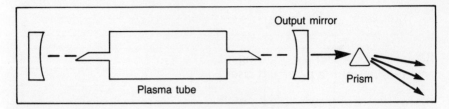

Fig. 14.15   *If resonator feedback is unrestricted, an ion laser will oscillate on several lines simultaneously*

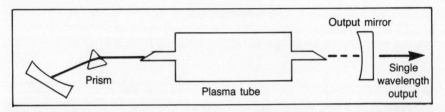

Fig. 14.16   *Oscillation can be limited to a single wavelength by restricting the feedback*

A consequence of the high-current density in an ion-laser plasma tube is a physical pumping of the ions toward the the cathode, where they are neutralized. If a return path for these neutral atoms is not provided, they tend to build up at the cathode and extinguish the discharge. Different laser manufacturers have different designs for this return path; but it is usually a convoluted path from cathode to anode, allowing the atoms to migrate back slowly but preventing the discharge from taking that route.

To maximize the current density in the plasma tube, most ion lasers include a solenoid wrapped around the tube that generates a strong longitudinal magnetic field in the discharge. The electrons and ions spiral about the lines of force of this field and are thereby held near the center of the discharge rather than allowed to collide with the walls of the plasma tube.

QUESTIONS

1. A typical 2-mW HeNe laser might draw a 5-milliamp current at 1,600 v. Calculate the efficiency of the laser. (Hint: Power consumed is equal to the product of current and voltage.)

2. How many modes of vibration are possible in a two-atom molecule such as H_2?

3. Explain why thermal birefringence of a solid-state laser rod is a loss mechanism in a polarized laser.

4. Explain why the wavelength of the laser that pumps a dye laser must be longer than the wavelength of the dye laser itself.

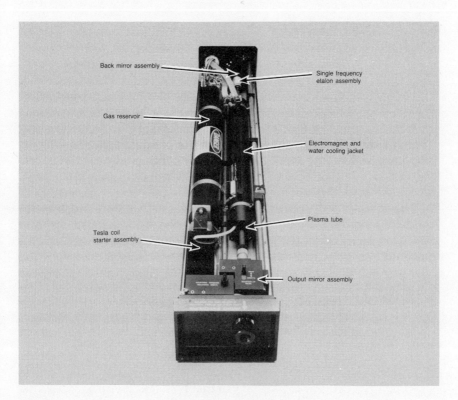

Plate 14.3   *Interior view of an ion laser* (Courtesy Cooper LaserSonics)

Chapter

≡Fifteen

Holography

Holography is a photographic technique that creates images which are much more realistic than the photographs produced by an ordinary camera. The basic difference between holography and conventional photography is that holography records the *phase* of light as well as its amplitude, while conventional photography records only the amplitude of light. As we'll see, this difference leads to some remarkable characteristics of holographic images.

This chapter will explain how a hologram is created and how it's used to produce a holographic image. Then we'll explore some of the commercial applications of holography for inspection and quality control.

How holography works

Fig. 15.1 shows a subject being illuminated by coherent light from a laser. An observer sees the subject when the light reflected from the subject is focused by the lens of his eye, forming an image on his retina. Likewise, an image of the subject could be formed on the film in a camera. The resulting photograph would be one way of recording the information—or at least part of the information—contained in the

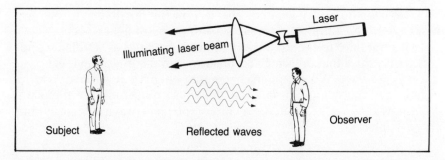

Fig. 15.1 *The observer "sees" the subject when light waves reflected from the subject reaches his eyes*

reflected light. Specifically, the photograph would record information about the amplitude of the light waves reflected from the subject. When examined, the photograph would show an image of the object as it would appear from the camera's viewpoint.

A hologram, on the other hand, records all the information contained in the light reflected from the subject. When the subject is illuminated, the hologram will reproduce the light waves that reflected from the object exactly as they were when the object was in place. Fig. 15.2 shows the hologram illuminated from behind by a laser. The light waves traveling toward the observer are exactly the same as the light waves reflected from the object when it was there. The observer sees the object in its full three-dimensional reality; he can even see different parts of the object by moving his head to look at it from different angles.

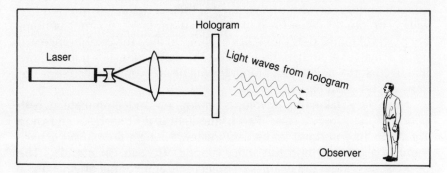

Fig. 15.2 *If a hologram is illuminated with a laser, the light waves reaching the observer's eye are exactly the same as if the object were still in place*

How is a hologram created? That is, how is it possible to record and later reproduce both the amplitude and phase of light waves?

It's not hard to record the amplitude of light waves. Ordinary photography does that. In Fig. 15.3 a piece of photographic film is exposed to a plane wave. What happens to the film when it's developed? It turns dark, and the degree of darkness depends on how intense the plane wave was. So we've created a record of the plane wave's intensity, but not its phase.

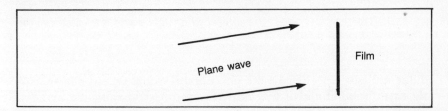

Fig. 15.3   *If a plane wave illuminates photographic film, the film will turn dark, recording the amplitude of the plane wave*

To record the wave's phase, we must use optical interference to create a hologram of the plane wave, as shown in Fig. 15.4. Here, a beam splitter and mirror have been added to the arrangement of Fig. 15.3. Part of the incoming wave has been split off, reflected from the mirror, and recombined with the rest of the wave at the film where interference takes place. Interference takes place because there is a path difference for the two rays of light striking point c (and every other point) on the film. Light can go from a to b to c, or it can go directly from a to c. If the length difference between these two paths is an integral number of whole wavelengths, then constructive interference will occur. On the other hand, if the path difference is an integral number of wavelengths plus one-half wavelength, then the interference will be destructive. A similar argument can be made for every point on the film, so that the developed hologram will be an evenly spaced series of lines as shown in Fig. 15.5.

Now, let's take this developed hologram and illuminate it with coherent light from a laser. Part of the light is diffracted, as shown in Fig. 15.5b (If you don't know why, review Young's double-slit experiment in Chapter 4 and the acousto-optic effect in Chapter 11.) The point is, this diffracted beam is an exact replication of the original plane wave. Both its amplitude and phase have been recorded in the hologram and are reproduced in the diffracted beam.

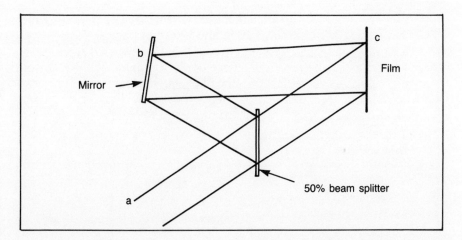

Fig. 15.4 *Arrangement for recording the hologram of a plane wave*

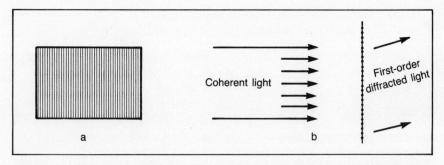

Fig. 15.5 *The interference pattern, or hologram, created on the film in Fig. 15.4 will be a series of evenly spaced straight lines (a). When the hologram is illuminated with coherent light (b), it recreates the original plane wave*

Now it's not really very exciting to make a hologram of a plane wave, so the next thing to do is to make a hologram with the light reflected from an object. The simplest of all possible objects is a point. Fig. 15.6a shows how this hologram is produced.

As before, part of the incoming laser beam is split off and passed directly to the film where it interferes with the other light—in this case, light reflected from the object. When the hologram is developed, it will consist of a series of irregularly spaced bars. And when this hologram is illuminated by coherent light in Fig. 15.6b, the spherical waves reflected from the point in Fig. 15.6a will be reproduced.

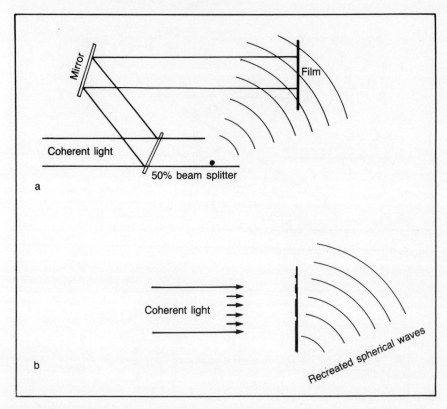

Coherent light

50% beam splitter

a

Coherent light

Recreated spherical waves

b

Fig. 15.6 *Arrangement for creating a hologram of a point (a). When the holo-gram is illuminated with coherent light (b), it reproduces the light scattered from the point, even though the object (the point) has been removed*

Now we can take the final step of creating a hologram of a real, extended object because such an object is simply a collection of many points. In Fig. 15.7, light reflected from each illuminated point on the object interferes with light reflected to the film by the beam splitter and mirror, creating an incredibly complex interference pattern on the film. (The pattern is too fine to see with the unaided eye; when you look at a hologram in ordinary light, you'll see patterns of circles and swirls that are incidental to the hologram itself.) But when this hologram is illu-minated with coherent light, as in Fig. 15.8, all the waves reflected from the object are recreated just as they existed in the first place. As far as the observer can tell, he is looking at the object itself because the light waves reaching his eyes are exactly the same as those reflected from the object.

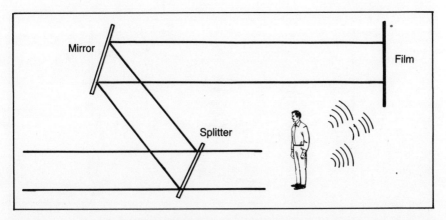

Fig. 15.7 *Creating a hologram of a real object involves recording the light scattered from every point of the object.*

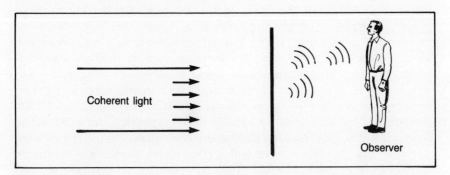

Fig. 15.8 *The light scattered from the object is recreated when the hologram is illuminated with coherent light.*

What we've seen here is the basic requirement for creating a hologram: The incoming light is split into a *reference beam* that goes directly to the film and an *object beam* that is reflected from the object to the film. Interference between these two beams at the film creates the hologram. As long as provision is made for this interference to take place, any arrangement of the equipment will work. The image produced by the hologram in Fig. 15.7 will be backlighted. An arrangement that would produce a frontlighted image is shown in Fig. 15.9.

Industrial holographic applications

Holographic inspection systems are found on many commercial assembly lines where they routinely locate defects invisible to human

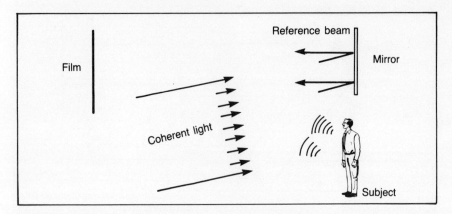

Fig. 15.9   *Another arrangement for creating a hologram of a subject. In this case, the subject would appear lighted from the front*

inspectors. Although there are many different approaches to holographic inspection, all use the same basic technique: Two holograms are created of the object, and these two holograms are overlaid and viewed simultaneously. Any change in the object that occurred between the two holograms shows up as interference fringes. For example, Fig. 15.10 shows a piece of material with an internal defect. When the material is stressed, the surface is distorted slightly. In Fig. 15.10b the stress could be caused by a clamp that compressed the width of the block. If a hologram is made before and after stressing and the two are overlaid, interference fringes will appear along the surface distortion caused by the internal defect. This distortion could not be detected visually, for it would only be micrometers or less in dimension. This type of holographic inspection is often used to locate defective aircraft tires or delaminations in layered materials.

Holography can also be used to analyze the vibrations of mechanical parts. For example, the read/write head of a computer's disk memory is moved across the disk and is stopped very rapidly. These sudden, jerky movements can generate catastrophic vibrations if the head is not designed properly. Holography can be used as a tool to analyze the vibrational modes of the head, as shown in Fig. 15.11. In this case two holograms were taken in rapid sequence, and the interference fringes created when the two were overlaid show the flexing of the head. The fringes across the arm indicate a bending vibrational mode, and the absence of fringes along the direction of the arm indicates there is little torsional motion.

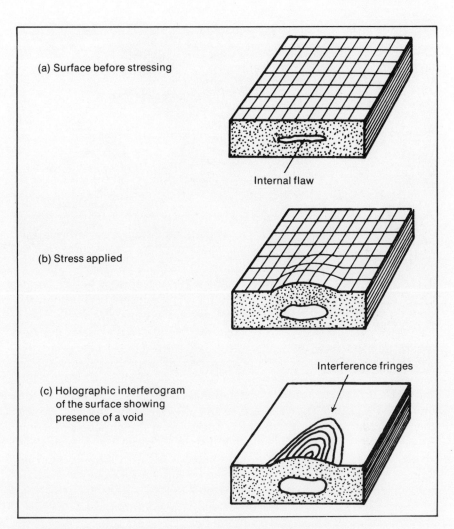

(a) Surface before stressing

Internal flaw

(b) Stress applied

Interference fringes

(c) Holographic interferogram
of the surface showing
presence of a void

Fig. 15.10 *Detection of a void by holographic interferometry.*

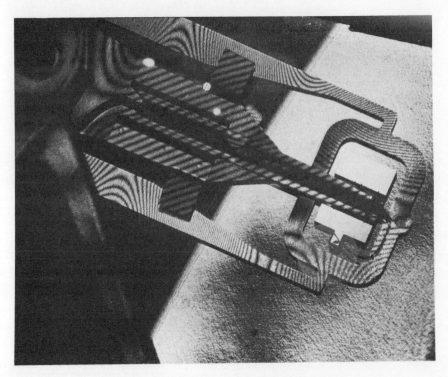

Fig. 15.11   *Holographic interferometry of a computer disk-read head.*

Chapter

Sixteen

Spectroscopy

Laser spectroscopy is an enormous field with literally hundreds of books and monographs detailing its intricacies. There are several international conferences every year devoted solely to the subject, and many of the world's most brilliant physicists and chemists have made laser spectroscopy their speciality. Needless to say, we will merely scratch the surface of the subject in this chapter.

Nonetheless, in many ways this chapter is one of the most fascinating in the book. We'll begin by learning what laser spectroscopy is—and why it's hard to do. Then we'll look at two examples of modern laser spectroscopy: Doppler-free spectroscopy and Raman spectroscopy. If you understand the techniques discussed here, you'll appreciate the ingenuity that went into their development.

What is laser spectroscopy?

In its simplest form, spectroscopy is the science of mapping out the energy levels of atoms and molecules. In more complex forms, it can be a tool for inferring the physical and chemical status of a sample under study. Although spectroscopy is conceptually simple, it can

207

quickly become very complex because even the simplest atoms have hundreds of energy levels.

A simple laser spectroscopy experiment is diagramed in Fig. 16.1. A tunable laser shines through the sample cell, and the transmission is recorded as a function of the laser's wavelength. Suppose absorption is observed at two different wavelengths, as shown in the figure. What can you conclude about the energy levels of the material in the sample cell?

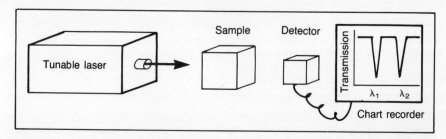

Fig. 16.1   *As the wavelength of the laser is tuned, the sample absorbs light at two distinct wavelengths*

The answer is shown in Fig. 16.2. You can conclude that there are two energy levels in the sample, and that their energy is greater than the energy of the ground state by an amount equal to the energy of the absorbed photons.

Pretty simple, isn't it? So what makes spectroscopy so hard?

There are several things. One thing that we will examine in this chapter is Doppler broadening. Suppose you're trying to identify two closely spaced energy levels in a gas. The energy levels might look like

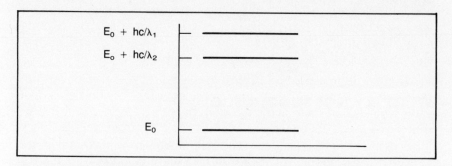

Fig. 16.2   *The data taken in the experiment in Fig. 16.1 indicate the sample has these energy levels*

those in Fig. 16.3. Now, if all the atoms in the gas stood still and if the bandwidth of your laser were narrow enough, what you would see on the strip-chart recorder of Fig. 16.1 is shown in Fig. 16.4a. But as you know, atoms in a gas don't stand still; they move around rapidly, bouncing off each other and off the walls of the container. And when they're moving toward or away from the laser, their absorption is Doppler shifted. (In Chapter 10 we saw that the emission of moving atoms in a gas laser was Doppler shifted. The same principle is operating here for absorption.)

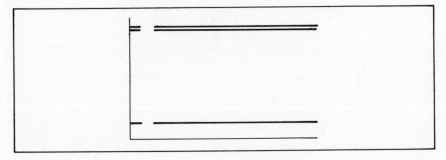

Fig. 16.3 *Two closely spaced energy levels can be difficult to resolve, as shown in Fig. 16.4*

So some of the atoms in the sample cell will have an absorption Doppler shifted as shown in Fig. 16.4b while others—those moving in the opposite direction—will be shifted as shown in Fig. 16.4c. And what you'll observe will be the sum of all these absorptions, as shown in Fig. 16.4d. The two distinct transitions have been smeared out into one big lump! So this experiment can't tell if there's one energy level or two (or more, for that matter) in the sample.

Another frequent problem in spectroscopy is weak signals, signals that originate from transitions that absorb or emit very small amounts of light. These weak signals can often be lost in the noise of the experiment. In Fig. 16.5, the upper trace represents a weak signal and the lower trace represents the normal background noise of the experiment—room lights, thermal noise in the detector, electrical pickup in cables, etc. When these two are added, the signal is completely lost. So the problem is how to extract a small signal from a noisy background.

Other complications that make laser spectroscopy such an interesting field include the need to study very fast transitions (subpicosecond modelocked lasers are used), the need to resolve different transitions

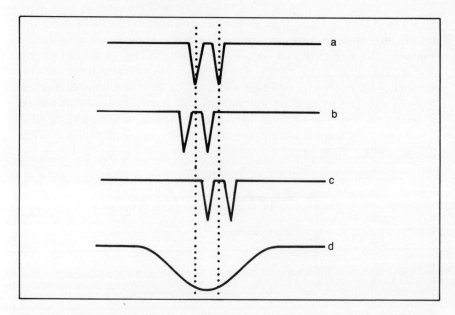

Fig. 16.4   *Absorption by the energy levels of Fig. 16.3 will be Doppler-shifted to (b) shorter wavelength or (c) longer wavelengths, or (a) unshifted, depending on the motion of the individual atom. But the total absorption from all atoms in the cell will look like (d)*

at the same wavelength, and the need to conduct experiments in hostile environments, like inside an internal-combustion engine or a coal-gasification plant.

Doppler-free spectroscopy

As we explained in the last section, Doppler broadening can be a severe limitation on the resolution of laser spectroscopy. Two close-lying energy levels can be smeared together to look like a single level. There are several direct approaches to solving this problem. Doppler broadening is caused by the motion of the atoms or molecules being studied, so the first thing you might do is slow down that motion: remove thermal energy from the sample. This works, up to a point and with some practical drawbacks. But the velocities of the atoms or molecules are proportional to the square root of temperature, so a big temperature reduction causes only a moderate reduction of thermal motion. And as a practical problem, chilling the sample can lead to bothersome condensation on the windows of the sample cell.

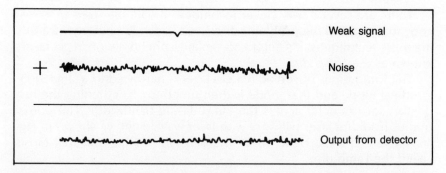

Fig. 16.5   *If signals are weak, they can be masked by noise generated in the experiment*

If you think about it a second, you realize that it isn't simply the motion of the atoms or molecules that causes Doppler broadening; it's all the different velocities. If all the atoms or molecules were moving with the same velocity (in the same direction), there would be no Doppler broadening. That's the reason for doing spectroscopy with atomic or molecular beams, as shown in Fig. 16.6. Since all the atoms here are moving with the same velocity, they all experience the same Doppler shift and there is no Doppler broadening.

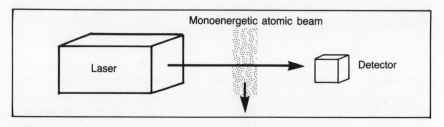

Fig. 16.6   *Because the atoms move perpendicular to laser beam, their absorption is not Doppler-broadened*

Molecular or atomic beam spectroscopy works fine, too, up to a point. It is impossible to create a perfectly monoenergetic molecular or atomic beam (for the same reason it's impossible to create a perfectly monochromatic laser: the uncertainty principle). It is difficult to prepare even a reasonably monoenergetic beam. If the material you're working with is rare or expensive, you're even worse off because beam experiments usually require a lot of sample material. And it is very hard to study short-lived states this way because they decay before they get to the laser beam.

There are several very clever techniques of Doppler-free spectroscopy that avoid all the problems discussed so far. But before introducing these techniques, it's helpful to explain a phenomenon in gas lasers known as the *Lamb dip*.

The Lamb dip is observed if a gas laser is narrowed to a single longitudinal mode and that mode is then tuned (by lengthening the resonator, for example) across the entire lasing bandwidth. The power output from the laser will vary with laser wavelength as shown in Fig. 16.7. There is a dip right where the peak should be, and that dip is called the Lamb dip.

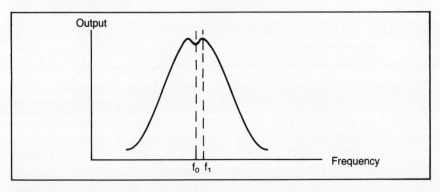

Fig. 16.7 *The Lamb dip is observed if a single-frequency gas laser is tuned across its gain bandwidth*

Understanding the reason for that dip is essential for understanding the Doppler-free spectroscopies we are about to discuss. The actual Doppler-broadened gain of the laser has no such dip in the middle, of course, but the dip is there because light moves in both directions inside a laser resonator. It might be helpful to review the discussion of Doppler broadening in Chapter 10 if you don't remember it well.

To see where the Lamb dip comes from, think about what happens when the laser is tuned to f_1, slightly off its center frequency in Fig. 16.7. There are two groups of atoms in the laser that can contribute to the laser's output: atoms moving with a certain velocity toward the output mirror and atoms moving with that same velocity away from the output mirror. Why? This is where it gets tricky. Let's think about the group moving toward the output mirror. They can emit light at an up-shifted frequency toward the mirror and at a down-shifted frequency away from the mirror. (A forward-moving train's whistle will sound up-shifted to someone standing in front of the train and down-shifted to some standing behind the train.) But the laser is tuned to provide feedback only at the up-shifted frequency, so our atoms will be stimulated to emit only at the up-shifted frequency.

The same argument holds for the group of atoms moving with the same velocity away from the output mirror. They will be stimulated to emit at an up-shifted frequency—f_1—in a direction away from the output mirror. But there is circulating power inside the resonator moving in both directions. So one way or the other, both of these groups of atoms—those moving toward the output mirror with a certain velocity and those moving away with the same velocity—will add to the circulating power in the laser at frequency f_1. Other atoms moving at other velocties will not be stimulated to emit at all.

Now, what happens when the laser is tuned exactly to its center frequency, f_0? There is now only one group of atoms that can contribute: those atoms standing still (or moving exactly perpendicular to the resonator axis). So there are fewer atoms that can contribute to the circulating power when the laser is tuned precisely to its line center, and there is a dip in output power at line center.

If you understand the origin of the Lamb dip, you're ready to understand a powerful Doppler-free technique called *saturation spectroscopy*. The experimental arrangement is diagramed in Fig. 16.8. A beam splitter divides the power in a narrow-bandwidth laserbeam unevenly between a powerful "saturating" beam and a weak "probe" beam. These two beams are then passed through the gaseous sample in opposite directions, and the power in the probe beam that is transmitted through the sample is measured with a detector.

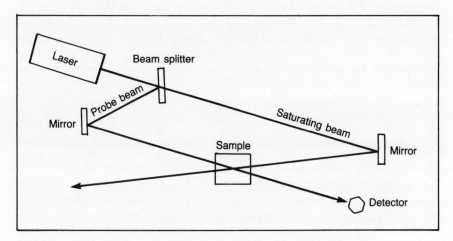

Fig. 16.8 *Experimental arrangement for saturation spectroscopy*

Fig. 16.9 shows the observable Doppler-broaded absorption spectrum of the sample and the real absorption of the atoms without Doppler broadening. That is, if you could examine just the nonmoving atoms in the sample and ignore all the others, you would see the narrow absorption in Fig. 16.9. But ordinary spectroscopy, like the experiment diagramed in Fig. 16.1, forces you to look at all the atoms, so you see the wide absorption in Fig. 16.9. The advantage of saturation spectroscopy is that it does, in fact, allow you to look at only the nonmoving atoms (or those moving perpendicular to the beams, which make no contribution to the Doppler broadening).

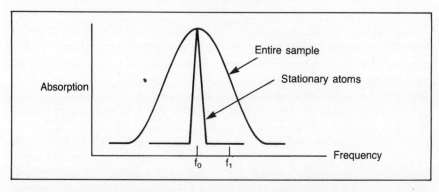

Fig. 16.9   *The observable absorption spectrum of all molecules in a gas sample and the absorption spectrum of only the stationary molecules*

To see how saturation spectroscopy works, think about what happens when the laser, whose bandwidth is as small or smaller than the non-Doppler-broadened absorption linewidth, is tuned slightly off the line center at $f_1 = f_0 + \Delta f$ in Fig. 16.9. Inside the sample cell there are some atoms moving to the left with just the right velocity so that they can absorb light at this frequency from the saturating beam. One of these atoms is shown in Fig. 16.10.[1] If the saturating beam is powerful

---

[1]*In discussing the Lamb dip, we noted that when a source is moving toward an absorber (e.g., when a train whistle is moving toward an ear), the emitted frequency appears to the absorber to be shifted up. On the other hand, if the absorber can only absorb a certain frequency, then as it and the source move toward each other, the source must be shifted down in frequency for absorption to occur. If Fig. 16.10 showed an emitting atom rather than an absorbing one, the sign of the frequency shift would be opposite.*

enough, it can completely saturate the transition. That is, if there are a lot more photons in the saturating beam than there are atoms in the absorbing velocity group, the photons can overwhelm the atoms, creating as many atoms in the upper level of the transition as in the lower level. This means that a photon in the saturating beam has a 50–50 chance of being absorbed and a 50–50 chance of stimulating emission and creating another photon. On average, though, as many photons will come out as went in when the transition is saturated; the absorption has become bleached.

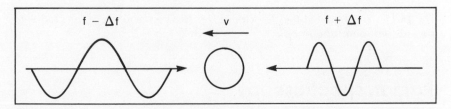

Fig. 16.10 *An atom moving to the left absorbs from the right at an up-shifted frequency and from the left at a down-shifted frequency*

And what's happening to the probe beam while all this is going on? Well, not much. The atoms that have been saturated by the saturating beam can't absorb light from the probe beam. These saturated atoms absorb light coming from the left at $f - \Delta f$, as shown in Fig. 16.10, and the frequency of the probe beam is $f + \Delta f$. So the probe beam is absorbed by another group of atoms, namely those moving to the right with exactly the correct velocity. The weak probe beam doesn't have enough photons to saturate the transition in these atoms.

Now let's tune the laser to the center frequency, f_0 in Fig. 16.9. This time, the atoms standing still are saturated by the powerful saturating beam. But that's the same group of atoms that absorbs the probe beam. When the laser is tuned to within the Doppler-free bandwidth of the absorption, the saturating and probe beams interact with the same group of atoms. And because the saturating beam bleaches the absorption of these atoms, they can no longer absorb the probe beam. The probe beam passes unhindered to the detector.

So when you perform saturation spectroscopy, you take data showing the signal at the detector as a function of laser frequency, as shown in Fig. 16.11. But in fact the curve that results is exactly the Doppler-free absorption of the sample.

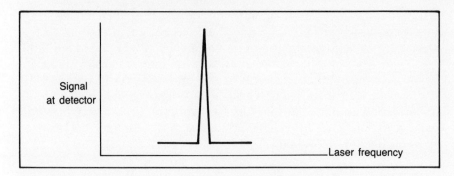

Fig. 16.11 *The signal at the detector in Fig. 16.8 corresponds to the Doppler-free absorption of the sample*

Raman spectroscopy

Raman spectroscopy is an effective tool for studying the vibrational and rotational energy levels of molecules that are not observable by the usual absorption and emission techniques. Raman scattering involves a new effect in our discussion of light: part of the energy in a photon is transferred to (or from) the vibrational/rotational energy of a molecule.

Before discussing Raman spectroscopy, let's take a little time to learn about simple Raman scattering, the effect exploited in Raman spectroscopy. Recall from Chapter 6 that molecules have vibrational and rotational energy levels. Just as photons are the particles associated with the quantization of light waves, *phonons* are the particles associated with the quantization of vibrational or rotational energy. In Raman scattering, vibrational or rotational phonons are created when part of the energy in a photon passing through the sample is converted to vibrational or rotational energy of the sample. (Alternatively, phonons can be annihilated when their energy is added to that of a photon passing through the sample.)

Fig. 16.12 shows the concept of Raman scattering. The energy in the input photon is divided between the output photon and a phonon that is created in the medium. Normally, only a small portion of the incoming photon's energy is passed to the phonon, so the output photon has a wavelength only slightly longer than that of the input photon.

When we talk about phonons being created from the energy in a photon, we're talking about the *Stokes* component of Raman scatter-

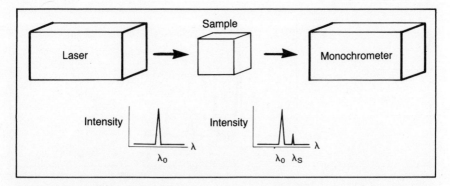

Fig. 16.12 *Spontaneous Raman scattering. Some of the photons passing through the sample lose part of their energy to vibrational/rotational modes of the sample*

ing. In this case the output wavelength is longer than the input wavelength because the photon loses energy to the phonon. But if there's already a (thermal) phonon in the scattering material, its energy can be added to that of a photon, in which case the wavelength of the scattered light is shorter than that of the input light. The *anti-Stokes* component of Raman scattering is generally much weaker than the Stokes component because there are so few thermal phonons from which the light can scatter.

To make things even more complicated, a photon can interact with more than one molecule as it passes through the scattering medium. If it's scattered from two molecules, losing energy and creating a phonon each time, its wavelength shift would be twice as much as normal. This light is the *second Stokes* component of Raman scattering. So a typical Raman-scattering experiment, as diagrammed in Fig. 16.12, can produce a whole spectrum of scattered light, as shown in Fig. 16.13.

Now, lets' take a look at what a simple spontaneous Raman experiment would look like, as diagrammed in Fig. 16.12. The laser irradiates the sample with monochromatic light at wavelength λ_0. But a spectral examination of the light transmitted through the crystal reveals that a component at λ_S has been added. Remember that the goal of any experiment is to study the energy level of the sample. What can you deduce from the Raman spectrum about the energy-level diagram of the sample?

The answer is shown in Fig. 16.14. This diagram shows the ground vibrational level (level 0), the first excited vibrational level (level 1), and

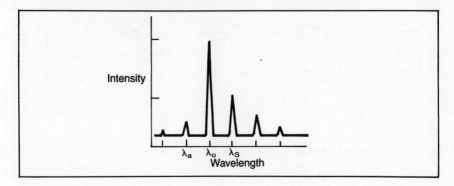

Fig. 16.13 *The light Raman scattered from a single vibrational mode has several Stokes and anti-Stokes components*

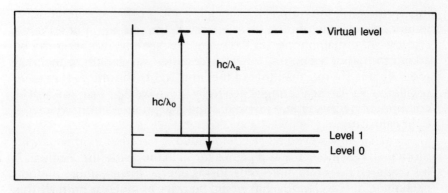

Fig. 16.14 *Two vibrational levels in the sample of Fig. 16.12. When Raman scattering occurs, the sample simultaneously absorbs a photon of energy hc/λ_o and emits a photon of energy hc/λ_s.*

a "virtual" level. This virtual level is not a real energy level, and the molecule cannot exist in this level. You can think of the virtual level as merely a convenience in drawing the Raman-scattering process on the energy-level diagram.

In Raman scattering, the molecule instantaneously absorbs a photon with energy hc/λ_0 and re-emits a photon with energy hc/λ_S. The extra energy (hc/λ_0 − hc/λ_S) is left in the vibrational mode of the molecule. This process is represented by the arrow up (absorption) and down (emission) in Fig. 16.14.

But from a measurement of the wavelength shift in Fig. 16.12, you can calculate the energy of the first vibrational level in Fig. 16.14. This

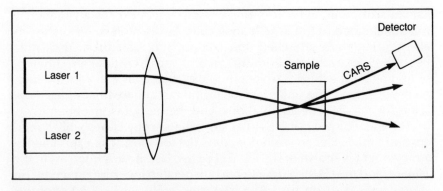

Fig. 16.15 *Experimental arrangement for coherent anti-Stokes Raman scattering (CARS)*

was precisely our goal in the experiment: to deduce what the energy-level diagram of the molecule looks like. In a real Raman scattering experiment, the spectrum of the scattered light would be more complicated than shown in Fig. 16.12 because there might be components due to scattering from energy levels above the first excited vibrational state. Another complication to the simplified spectrum in Fig. 16.12 would arise if the resolution were great enough to see the rotational substructure of the vibrational transition. Remember that the energy-level diagram in Fig. 16.14 is a simplification that does not show rotational levels. If you're not clear on how the rotational levels would affect the spectrum of scattered light, you should go back and review the effect of rotational levels in a CO_2 laser in Chapter 14.

Even if the incoming light were scattered by only one vibrational level without rotational substructure, the spectrum would not really be as simple as the one shown in Fig. 16.12. As we learned earlier, a single photon can interact with more than one molecule, so higher-order Stokes and anti-Stokes components can be present in the scattered light, as shown in Fig. 16.13.

So far, we've been talking about spontaneous Raman scattering and its spectroscopic applications, but there are several types of coherent Raman scattering that can be observed with powerful lasers. These coherent Raman techniques can often be used to study the spectroscopy of vibrational/rotational levels that cannot be detected via spontaneous Raman scattering.

The practical difference between spontaneous Raman spectroscopy and coherent Raman spectroscopy is that the coherent approach requires an extra laser tuned to a different wavelength. In coherent

Raman scattering, light from one or both lasers excites a vibrational/ rotational mode of the sample molecule. As the molecule vibrates, its electric fields are stretched and compressed, and these oscillating fields interact with light from the second laser to produce a coherent output beam at a third wavelength. (In spontaneous Raman scattering, on the one hand, the single laserbeam interacts only with thermal—not laser-driven—molecular vibrations, and the output is incoherent.)

The best-known coherent Raman spectroscopy is probably coherent anti-Stokes Raman scattering, or CARS for short. The experimental arrangement is shown in Fig. 16.15. The two laserbeams intersect in the sample, and the CARS output beam is generated in a third direction. An energy-level diagram for CARS is shown in Fig. 16.16. Two photons, one from each laser, drive a vibrational mode if the energy difference between the two photons is equal to the energy of the mode. This process is shown by the two arrows on the left side of Fig. 16.16. Then another laser photon is scattered from the vibrating molecule, absorbing the energy of its vibration, as shown by the other two arrows in the figure.

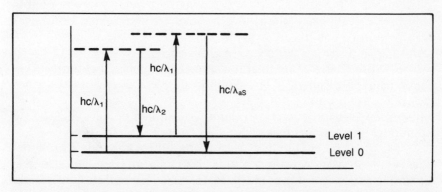

Fig. 16.16 *In CARS, two laser photons drive a vibrational mode that scatters a third laser photon*

So the experiment shown in Fig. 16.16 works this way: One or both lasers are tuned until a CARS signal is observed at the detector. When that happens, there is a vibrational mode in the sample molecule whose energy is exactly equal to the energy difference between photons from the two lasers. And remember that this is the goal of spectroscopy: to deduce what the energy-level diagram looks like.

What are the advantages of CARS and other coherent Raman spectroscopic techniques over spontaneous Raman spectroscopy? A big advantage of coherent Raman spectroscopy is that the output signal is generated in a beam rather than randomly generated in all directions as it is in spontaneous scattering. That means that all the light can easily be collected with a single detector. Moreover, the efficiency of most coherent Raman techniques is far greater than the corresponding efficiency for spontaneous Raman scattering. CARS efficiencies, in particular, can be 10^5 higher than spontaneous Raman scattering efficiency.

Questions

1. Can you think of a way to use spontaneous Raman scattering to measure the temperature of a sample? (Hint: Why are the anti-Stokes components weaker than the Stokes components?)

For further reading

Eisberg. *Fundamentals of Modern Physics*. New York: John Wiley & Sons, 1961.

Fowles. *Introduction to Modern Optics*. New York: Holt-Reinhard-Winston, 1976.

Hall and Carlsten. *Laser Spectroscopy III*. New York: Springer-Verlag, 1977.

Hecht and Teresi. *Laser, Supertool of the 1980s*. New York: Ticknor & Fields, 1982.

Jenkins and White. *Fundamentals of Optics*. New York: McGraw-Hill, 1976.

Koechner. *Solid-State Laser Engineering*. New York: Springer-Verlag, 1976.

Leith and Upatnieks. "Photography by Laser." *Scientific American*. June 1965, p. 24.

O'Shea, Callen, and Rhodes. *An Introduction to Lasers and Their Applications*. Reading, Massachusetts: Addison Wesley, 1977.

Pennington. "Advances in Holography." *Scientific American*. February 1968, p. 40.

National Geographic. March 1984 issue devoted to lasers and holography.

Saunders et al. *Lasers*. New York: McGraw-Hill, 1980.

Siegman. *An Introduction to Lasers and Masers*. New York: McGraw-Hill, 1971.

Verdeyen. *Laser Electronics*. Englewood Cliffs, New Jersey: Prentice-Hall, 1981.

Yariv. *Quantum Electronics*. New York: John Wiley & Sons, 1967.

Index